T0394945

Wind Energy and Wildlife Impacts

Regina Bispo • Joana Bernardino • Helena Coelho
José Lino Costa
Editors

Wind Energy and Wildlife Impacts

Balancing Energy Sustainability with Wildlife Conservation

Editors

Regina Bispo
Department of Mathematics and Centre for Mathematics and Applications (CMA)
Faculty of Sciences and Technology of NOVA University of Lisbon
Monte da Caparica, Portugal

MARE – Marine and Environmental Sciences Centre
Faculty of Sciences of the University of Lisbon
Lisbon, Portugal

Helena Coelho
Bioinsight
Odivelas, Portugal

Joana Bernardino
CIBIO/InBIO – Research Centre in Biodiversity and Genetic Resources
University of Porto
Vairão, Portugal

José Lino Costa
MARE – Marine and Environmental Sciences Centre
Faculty of Sciences of the University of Lisbon
Lisbon, Portugal

ISBN 978-3-030-05519-6 ISBN 978-3-030-05520-2 (eBook)
https://doi.org/10.1007/978-3-030-05520-2

Library of Congress Control Number: 2019931842

This Springer imprint is published by the registered company Springer Nature Switzerland AG.
The registered company address is: Gewerbestrasse 11, 6330 Cham, Switzerland

Preface

We are very pleased to introduce the proceedings of the Fourth International Conference on Wind energy and Wildlife impacts (CWW 2017), held on 6–8 September 2017, at the Centro de Congressos do Estoril, Portugal.

Wind power can be used in many ways and has several known advantages. It is renewable, does not involve carbon emissions, minimizes overdependence on traditional sources of electricity, and has domestic potential, among other important gains. However, the potential for impacts on wildlife is still a reality with potential to affect populations of many species, due to habitat loss and fragmentation, disturbance, nonnative invasive species, and increased mortality by direct collision. Hence, today's challenge is still to maximize wind energy's benefits while minimizing the risk to wildlife.

Aiming toward this goal, the International Conference on Wind energy and Wildlife impacts (CWW) gathers experts from all over the world aiming to promote the international cooperation among academics, researchers, and professionals, which, over the years, have contributed to building knowledge on this topic.

CWW started in 2011 being held since then biennially. The event has been continuously growing, and at the conference, in 2017, held at Estoril, Portugal, there were 136 presentations, 19 exhibitors, and over 340 registered participants, from 30 different countries all over the world. Oral conference presentations were organized in 15 parallel sessions, covering subjects related to species behavior (off- and onshore), fatality assessment, species fatality and vulnerability, mitigation, impact monitoring and risk assessment, planning and policy, population impact modelling, ecosystems and holistic approaches, and tools and technology. Conference program also included two sessions devoted to poster presentations. These proceedings volume provide a record of what was presented at CWW 2017. It contains 13 papers covering the subjects from both oral and poster presentations.

Many people have contributed to the organization of CWW 2017. We are most grateful to all members of the Organizing Committee that worked thoroughly for the success of this event and of the Scientific Committee for their commitment in supporting scientifically the event. We are also grateful to all keynote speakers. Finally, we acknowledge all the participants on the panel discussion, as well to all speakers for their presentations and contributions to the conference.

Lisbon, Portugal Regina Bispo
Vairão, Portugal Joana Bernardino
Odivelas, Portugal Helena Coelho
Lisbon, Portugal José Lino Costa

Acknowledgments

We would like to thank all the people who participated in the CWW 2017 organization and very especially to those who collaborated in organizing the event or served as referees for the conference proceedings:

Amanda Hale, Texas Christian University, United States of America
Ana Luísa Rodrigues, Portugal
Ana Teresa Marques, cE3c – Centre for Ecology, Evolution and Environmental Changes, Portugal
Andrea Copping, Pacific Northwest National Laboratory, United States of America
Andrea Talaber, Iberdrola Renovables, Spain
Andrew Gill, Cranfield University, United Kingdom
António Sá da Costa, APREN – Associação de Energias Renováveis, Portugal
Aonghais Cook, British Trust for Ornithology, United Kingdom
Carl Donovan, DMP Statistical Solutions, United Kingdom
Cris Hein, Bat Conservation International, United States of America
Daniel Skambracks, KfW, Germany
Edward Arnett, TRCP, United States of America
Fabien Quétier, Biotope, France
Finlay Bennet, Marine Scotland Science, United Kingdom
Francisco Moreira, CIBIO – University of Porto, Portugal
Fränzi Korner-Nievergelt, oikostat GmbH, Switzerland
Glen Tyler, Scottish Natural Heritage, United Kingdom
Henrique Cabral, MARE, Portugal
Jan Olof Helldin, Calluna AB, Sweden
Jerry Roppe, Avangrid Renewables, United States of America
Johann Köppel, TU Berlin, Germany
Katharina Fließbach, Research and Technology Centre (FTZ), Kiel University, Germany
Manuela de Lucas, Doñana BS, Spain
Manuela Huso, USGS, United States of America

Miguel Mascarenhas, Bioinsight, Portugal
Miguel Repas, STRIX, Portugal
Nuno Matos, APAI/Matos, Fonseca & Associados, Portugal
Rachael Plunkett, SMRU Consulting, United Kingdom
Roel May, Norwegian Institute for Nature Research (NINA), Norway
Samantha Ralston-Paton, BirdLife South Africa, South Africa
Sofia Menéres, ISPA – Instituto Universitário, Portugal
Taber Allison, American Wind Wildlife Institute, United States of America
Teresa Simas, WavEc, Portugal
Timóteo Monteiro, EDP Renováveis, Portugal

Contents

Contributors

Efi Armeni Biology Department, University of Crete, Heraklion, Greece

Finlay Bennet Marine Scotland Science, Aberdeen, Scotland, UK

Juliane Biehl Environmental Assessment and Planning Research Group, Berlin Institute of Technology (TU Berlin), Berlin, Germany

Alexander Bittner Deutsche Bundesstiftung Umwelt, Osnabrück, Germany

Jan Blew BioConsult SH GmbH & Co. KG, Husum, Germany

Stuart Clough APEM Inc, Gainesville, FL, USA

Andrea Copping Pacific Northwest National Laboratory, Seattle, WA, USA

Jean-Claude Dauvin Laboratoire Morphodynamique Continentale et Côtière, CNRS, Normandie Univ, UNICAEN, UNIROUEN, Caen, France

Victoria Gartman Berlin Institute of Technology, Berlin, Germany

John Halley Department of Biological Applications & Technology, University of Ioannina, Ioannina, Epirus, Greece

Manuela Huso US Geological Survey, Corvallis, OR, USA

Erik Mandrup Jacobsen Orbicon A/S, Taastrup, Denmark

Flemming Pagh Jensen Orbicon A/S, Taastrup, Denmark

Dale M. Kikuchi Tokyo City University, Kanagawa, Japan

Wataru Kitamura Tokyo City University, Kanagawa, Japan

Ryosuke Komachi Graduate School of Science and Technology, Niigata University, Niigata, Japan

Johann Köppel Environmental Assessment and Planning Research Group, Berlin Institute of Technology (TU Berlin), Berlin, Germany

Gregory Lampman NYSERDA, Albany, NY, USA

Miguel Mascarenhas Bioinsight, Odivelas, Portugal

Roel May Norwegian Institute for Nature Research (NINA), Trondheim, Norway

Stephanie McGovern APEM Ltd, Stockport, UK

Sachiko Moriguchi Faculty of Agriculture, Niigata University, Niigata, Japan

Haruka Mukai Graduate School of Science and Technology, Niigata University, Niigata, Japan

Toru Nakahara Graduate School of Fisheries and Environmental Sciences, Nagasaki University, Nagasaki, Japan
Kitakyushu Museum of Natural History and Human History, Fukuoka, Japan

Tiago Neves Bioinsight, Odivelas, Portugal

Juha Niemi Signal Processing Laboratory, Tampere University of Technology, Pori, Finland

Stamatina Nikolopoulou Natural History Museum of Crete, University of Crete, Heraklion, Greece

Nathalie Niquil Unité Biologie des Organismes et Ecosystèmes Aquatiques (BOREA), MNHN, CNRS, IRD, Sorbonne Université, Université de Caen Normandie, Université des Antilles, Caen, France

Ann Pembroke Normandeau Associates Inc, Bedford, NH, USA

Jean-Philippe Pezy Laboratoire Morphodynamique Continentale et Côtière, CNRS, Normandie Univ, UNICAEN, UNIROUEN, Caen, France

Aurore Raoux Laboratoire Morphodynamique Continentale et Côtière, CNRS, Normandie Univ, UNICAEN, UNIROUEN, Caen, France

Mark Rehfisch APEM Ltd, Stockport, UK

Luís Rosa Bioinsight, Odivelas, Portugal

Tsuneo Sekijima Faculty of Agriculture, Niigata University, Niigata, Japan

Juha T. Tanttu Mathematics Laboratory, Tampere University of Technology, Pori, Finland

Diana Vieira Bioinsight, Odivelas, Portugal

Volker Wachendörfer Deutsche Bundesstiftung Umwelt, Osnabrück, Germany

Simon Warford APEM Ltd, Stockport, UK

Jessica Weber Environmental Assessment and Planning Research Group, Berlin Institute of Technology (TU Berlin), Berlin, Germany

Julia Robinson Wilmott Normandeau Associates Inc, Gainesville, FL, USA

Stavros M. Xirouchakis Natural History Museum of Crete, University of Crete, Heraklion, Greece
School of Science & Engineering, University of Crete, Heraklion, Greece

Noriyuki M. Yamaguchi Graduate School of Fisheries and Environmental Sciences, Nagasaki University, Nagasaki, Japan

The Role of Adaptive Management in the Wind Energy Industry

Andrea Copping, Victoria Gartman, Roel May, and Finlay Bennet

Abstract Adaptive management (AM) is a systematic process intended to improve policies and practices and reduce scientific uncertainty by learning from the outcome of management decisions. Although many nations are considering the use of AM for wind energy, its application in practice and in policy has been limited. Recent applications of AM have revealed fundamental differences in the definition of AM, its applications, and the projects or planning processes to which it might be applied. This chapter suggests the need for a common understanding and definition of and framework for AM and its application to wind energy. We discuss a definition of AM and technical guidance created by the United States (US) Department of the Interior's (DOI's) Adaptive Management Working Group. The chapter also examines how AM has been applied to wind energy development in several European nations and in the USA. The challenges and opportunities associated with implementation of AM for wind development are addressed, management actions in nations that exhibit attributes of AM are compared, and pathways to appropriate application and potential broader use of AM are explored.

Keywords Adaptive management · Wind energy development · Wind and wildlife interactions

A. Copping (✉)
Pacific Northwest National Laboratory, Seattle, WA, USA
e-mail: andrea.copping@pnnl.gov

V. Gartman
Berlin Institute of Technology, Berlin, Germany
e-mail: victoriagartman@clarkgroupllc.com

R. May
Norwegian Institute for Nature Research (NINA), Trondheim, Norway
e-mail: Roel.May@nina.no

F. Bennet
Marine Scotland Science, Aberdeen, Scotland, UK
e-mail: Finlay.Bennet@gov.scot

R. Bispo et al. (eds.), *Wind Energy and Wildlife Impacts*,
https://doi.org/10.1007/978-3-030-05520-2_1

1 Introduction

All sources of energy have impacts on the environment. Understanding these effects is critical to helping countries make informed decisions about the relative costs and benefits of various energy solutions. Rapid and large-scale development of renewable energy and other developments challenge our ability to anticipate, verify, and mitigate impacts on the environment [1]. Although land-based and offshore wind contribute significantly to the renewable energy portfolios of many nations, many uncertainties remain about the potential effects of wind farms on wildlife populations and the habitats that support them [2] with particular emphasis on species of concern due to depleted populations. Continued expansion of wind farms on land and at sea requires tools that allow for the environmental management of wind projects in the face of these uncertainties. One tool that shows promise for decreasing uncertainty and providing increased insight into the environmental effects of wind energy development is adaptive management (AM) [3].

AM has been discussed since the 1970s as a means of facilitating potential decision-making processes for addressing uncertainty, managing natural resources, and directing research [4]. AM has been used during the process of developing wind energy projects in the United States (USA) and is under consideration in other countries, because it has the potential to reduce scientific uncertainty and inform future wind energy project planning and management decisions [3]. However, AM has not been irrefutably shown to be a practical management tool, because of the lack of consistency in its definition, preferred outcomes, implementation practices, and scales of relevance. This lack of consistency has resulted in a wide range of outcomes and effectiveness for managing environmental uncertainties associated with the wind energy industry [5]. It is important to note that striving for consistency in the application of AM during all aspects of wind energy projects may not be desirable because the relevance of AM to mitigating environmental impacts is likely to be specific to each individual wind energy project.

This chapter discusses the science of AM from an international perspective and its intersection with policies and management practices that are common in the nations collaborating in WREN (Working Together to Resolve Environmental Effects of Wind Energy). The exact effects and interactions of AM with the regulations or policies of any individual country cannot be inferred from these discussions.

2 Adaptive Management as a Concept

2.1 Definition of Adaptive Management

Lack of a commonly accepted definition of AM presents a challenge for its implementation. We have chosen the definition used widely by management agencies in the USA, developed by the US Department of the Interior (DOI) Adaptive

Management Working Group's Adaptive Management Technical Guide [6] and companion Application Guide [7], referred to here as the DOI guidelines.

AM refers to a learning-based approach, or learning by doing, that leads to adaptations of management programs and practices based on what has been learned [4, 7]. The most widely accepted definition of AM comes from the US National Research Council [8] and has been adopted and further described in the DOI guidelines, as follows:

> Adaptive Management is a decision process that promotes flexible decision-making that can be adjusted in the face of uncertainties as outcomes from management actions and other events become better understood. Careful monitoring of these outcomes both advances scientific understanding and helps adjust policies or operations as part of an iterative learning process... [6, 8]

AM can be applied at the wind energy project scale, at which an AM approach is used to address scientific uncertainty and help inform management decisions related to specific projects (e.g., implementation of mitigation measures). Alternatively, AM can be applied at the planning scale using data and outcomes from individual and multiple projects to inform future regulations and development and management decisions [3]. The data collected may be similar for assessing scientific uncertainty and informing management decisions at both scales, but the spatial and temporal extent of monitoring data collection and analyses at the two scales may differ.

AM has been described as being passive or active [4, 9]. Passive AM describes the use of historical data to construct a single best estimate or model for response, by applying the best available science. Active AM uses more recent data to structure a range of alternative response models, and a management choice is made by taking into account the short-term performance and long-term value of knowing which alternative model best reflects the real-world situation.

Stakeholder engagement is an important additional ingredient that should be included throughout the AM process to generate initial research questions, review monitoring results, observe outcomes of management decisions, and ensure all affected individuals and organizations support AM objectives [6, 10].

2.2 *A Question-Driven Approach*

AM seeks to address scientific uncertainty and improve understanding of an environmental system using a question-driven, hypothesis-based approach. AM may not be suitable for application to all development projects. AM is not a trial-and-error process or a management approach that randomly implements alternative decisions if desired results are not achieved. Rather, AM maintains a hypothesis-based approach to meeting objectives agreed upon by the involved parties, and it typically uses quantitative or conceptual models to test hypotheses, provide management alternatives, and predict the consequences of management decisions. Post-installation monitoring of natural resource interactions provides data used to validate the models, address scientific uncertainties, and set a baseline for the managed environmental system [6, 8]. If a question-driven approach is not taken, the data collected may have limited relevance for use in validating models or supporting AM processes. A further set of challenges relates to the suitability of the experimental design for meaningfully addressing questions and the need to gather sufficient data to provide the levels of statistical power decision-makers require to implement AM.

2.3 Adaptability in the Face of the Uncertainty of Natural Variability

The AM process relies on maintaining a certain level of adaptability (flexibility) to ensure informed management decisions can be made in the face of uncertainty. As new information is gathered, management decisions may be amended to better accommodate the environmental system. Owing to the inherent natural variability of environmental systems and the inevitable measurement errors when measuring environmental interactions, uncertainty is a key attribute that must be accommodated by natural resource management. AM principles can be used to identify and understand natural variability and provide organizations with the knowledge and information to make informed management decisions in the face of uncertainty.

2.4 An Iterative Process

AM can be considered an iterative cycle, or "single-loop learning". As information and data are gathered over time, management approaches and decisions can be adapted to better accommodate the ecological process or system being managed, thereby leading to better understanding of the targeted ecological system and improved management decisions. A further purpose of AM relative to wind energy is to optimize the use of wind energy while maintaining environmental safeguards. In practice, AM should enable greater wind energy development if the associated environmental effects are shown to be insignificant.

As discussed by the DOI guidelines, AM also promotes "double-loop learning", or institutional learning. Figure 1 illustrates both single- and double-loop learning in the context of wind energy projects. The inner, single loop illustrates adjustments to individual projects based on data collected by that project (e.g., data to inform the use of mitigation measures at the project level). The outer, double-loop more likely takes place across multiple projects. It promotes the use of lessons learned from current and past projects to reconsider objectives and management alternatives and can potentially be used to inform future management decisions for other projects.

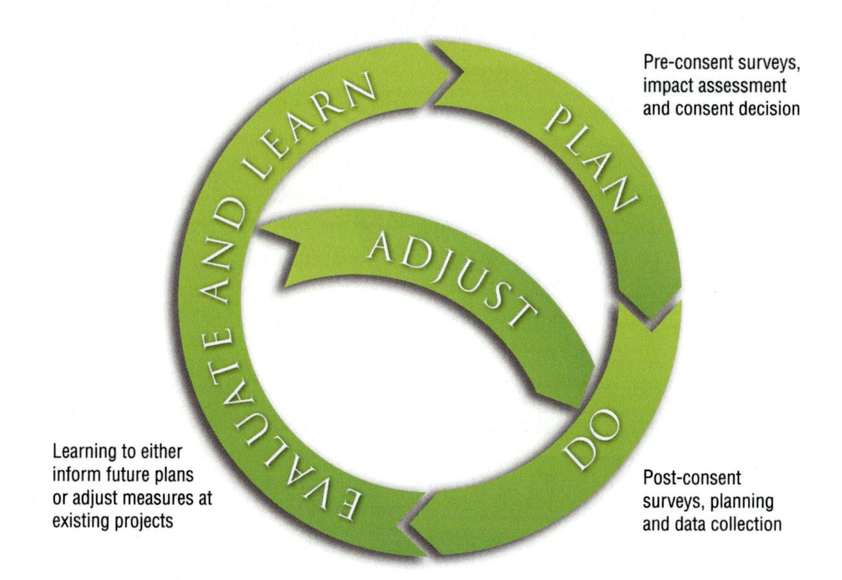

Fig. 1 Double-loop or institutional learning for AM. (Drawn based on Resilience to Transformation: The Adaptive Cycle [11])

2.5 Important Concepts Associated with AM

Several key concepts are critical for understanding the overarching objective of AM and how it is typically implemented.

- Scientific Uncertainty. The uncertainty that AM seeks to address is related to potential adverse outcomes associated with the development or operation of wind energy projects. However, AM may also be used to address the uncertainty associated with the overall effectiveness of certain mitigation measures or management decisions.
- Bounding. Bounding mitigation activities provides a level of financial security for project developers by limiting the amount of mitigation that may be required, while reassuring regulators that mitigation can be delivered if certain thresholds are surpassed.
- Scale. AM may be applied at the project scale to reduce the scientific uncertainty of an individual project or the effectiveness of particular mitigation measures (single-loop learning). AM may also be applied at the planning scale, where it takes place over multiple projects and may be used to collate information and monitoring data from different projects to inform the future management and permitting of wind energy projects (double-loop learning).

In practice, as mitigation activities are carried out, an AM approach could be used to evaluate the effectiveness of the mitigation actions, learn from experience, and reduce overall scientific uncertainty by informing more effective mitigation for use in future management decisions.

3 Application of Adaptive Management to Wind Energy Development

Each country and jurisdiction applies planning and management tools for the development of wind energy projects, based on the legal and regulatory system of that nation. Common tools or strategies applied to wind farm planning and management include the application of the precautionary principle [12, 13] and the mitigation hierarchy [14–17]. The application of AM principles to wind energy planning and management is not incompatible with these other management tools, but each must be carefully applied to ensure that a balance is struck between an overly restrictive approach that will not permit development of low-carbon energy production at all and one that protects key wildlife species and other environmental resources [5].

Although large volumes of post-installation data have been collected over two decades about interactions between wind turbines and wildlife for both land-based and offshore wind farms, many uncertainties remain along with a lack of understanding of basic processes that allow for effective mitigation of environmental effects. Similarly, monitoring data have not sufficiently informed future planning and management of wind farms over large spatial scales [18]. Often these data are considered to be "data rich and information poor", referred to as a DRIP condition [19]. These data may not yield sufficient information for a number of reasons, including the following:

- The data may not have been collected in a question-driven manner.
- The data utility may be undermined by issues of inadequate experimental design.
- The spatial scales over which monitoring has been carried out may not support the collection of samples to provide sufficient statistical power to meaningfully reduce scientific uncertainty.

AM can be applied at two scales for wind energy projects: the project scale and the planning scale. Data can be collected to reveal the impact of a single wind farm or a collection of projects on a population or segment of a population of concern. Because of the spatial and temporal scale of the data needed to understand large populations over wide geographic ranges, AM applied at the planning scale may be more effective at helping project planners understand how a wind energy project might affect a population of animals by using data from multiple projects over large geographic spaces. Conversely, the application of AM at the project scale may be limited for assessing population-level impacts, because project developers of single wind energy projects are unlikely to fund monitoring activities for an entire population that falls outside of the scope and range of their project, and the site-based results may have limited application to reducing uncertainty about the impacts on populations. Efforts are under way in the USA and other nations to broaden data collection to encompass areas larger than a single wind farm.

4 International Use of Adaptive Management

When reviewing international practices across a number of nations, the use of AM principles for wind energy projects was found to range from relatively frequent use in regulatory processes to no formal recognition or application of AM. AM is not a legal requirement in any of the countries reviewed. Case studies from Germany, the Netherlands, Norway, Portugal, Spain, Switzerland, the United Kingdom (UK), and the USA are described below (in alphabetic order).

4.1 Germany

In Germany, AM principles have been applied to several different projects. For example, the Ellern wind farm in Germany's southwest Rhineland-Palatinate attempted to mitigate the collision mortality of bats by curtailing turbine operation at wind speeds below 6 m/s from April to October. The mitigation was required locally, specified in the wind farm permit, and based on federal state guidelines. Data were collected during the first year of operation through carcass surveys and nacelle monitoring. After 1 year of operation, the monitoring data were compared with thresholds set by a group of stakeholders, including nature conservation organizations and the project proponent, and the curtailment methods were altered to ensure that the thresholds were met. Monitoring was only required for the first 2 years of wind farm operation, and subsequent adaptations to the monitoring plan were not intended [3].

Another wind farm located in North Rhine-Westphalia uses the cultivation of nearby farmland to trigger turbine shutdowns to avoid collisions with red kites (*Milvus milvus*) [20]. The birds are attracted to the newly mowed fields, presenting additional risk of collision with wind turbines. Shutdowns are required by permit under certain circumstances: during daytime periods if red kites nest within a 0.5 km radius of the turbines and for 3 days after cultivation activities. If monitoring of the surrounding area finds no nesting red kites within the minimum distance for 4 consecutive days, the measures are no longer required.

4.2 The Netherlands

AM principles have been used to adjust mandatory monitoring programs within projects for offshore wind farm projects. The offshore wind farm Luchterduinen includes intensive and regular contact between the competent authority and the wind developer to assess whether adjustment of the monitoring program is needed, based on monitoring results and information from other sources that have become available during the project [21]. Examples of major adjustments that have been made to the program include the addition of research on bats using bat detectors (which was not included in the monitoring program because the occurrence of bats at sea was largely unknown) and participation in the Disturbance Effects on the Harbour Porpoise Population in the North Sea project. The latter effort developed an individual-based model of the effects of underwater piling sound associated with installation of offshore wind turbines to support research into the effects of such sound on fish juveniles and larvae. These minor adjustments to the monitoring program led to a much more effective program than that originally scoped.

The use of AM principles among multiple offshore Dutch wind farms is becoming more common (double-loop learning). Currently, a third round of offshore wind farms is being planned in the Netherlands; each round builds on the knowledge acquired in the previous rounds to improve conditions and decrease constraints for offshore wind energy projects. The government implemented a new policy system for the third round; it improved implementation of AM by encouraging the selection of possible areas for offshore wind energy development, carrying out all preliminary environmental assessments, undertaking monitoring and research programs that will validate assumptions made in these assessments, and overseeing research into issues of financial importance to wind developers such as wind resource characterization, bathymetry, and sea bed characteristics [22]. Under this scheme, the government will draft decisions for each proposed wind farm site including all conditions and constraints for the development of a wind farm.

The third round consists of ten planned offshore wind farms of 350–380 MW capacity each. Wind farm site decisions will be drafted in five phases, one phase per two wind farms. Knowledge gathered during one phase will be applied in the next phase using an AM process. This process is new and just beginning to unfold, making it difficult to assess the extent to which AM will actually be applied. An evaluation at the end of 5 years will provide better insight into the questions about the use and efficacy of AM.

4.3 Norway

In Norway, the Nature Diversity Act could provide the legal mechanism for AM practices, because the precautionary and polluter pays principles are well established in the Act. However, in practice these principles are not used for wind energy and wildlife interactions because of the limited influence of the Environmental Agency in the consenting process. In addition, continuous tension exists between authorities and industry about who should be held responsible for financing monitoring and research on the environmental impacts of wind energy developments. There also is pressure to minimize total project costs in favor of profitability, which can act to compromise environmental considerations. Although

AM is currently not implemented in Norway, the renewable energy company Statkraft co-financed extensive research and monitoring at the Smøla wind farm from 2006 to 2017 [23, 24]. This effort included testing of mitigation measures in response to an official complaint from the Bern Convention to the Norwegian government concerning conflicts with white-tailed eagles (*Haliaeetus albicilla*). The investment has contributed to reducing the scientific uncertainty pertaining to both the extent of the impacts and effectiveness of mitigation measures.

4.4 Portugal

A good example of the principles of AM can be found at the Candeeiros wind farm located in the central portion of Portugal [25]. The Portuguese refer to it as an iterative approach to post-construction bird mortality monitoring. After 3 years of post-construction bird monitoring, the common kestrel (*Falco tinnunculus*) emerged as the species most commonly killed at the wind farm. As a result, the monitoring program was changed to study the kestrel population and evaluate the impact of the wind farm on this species. Although the common kestrel is not an endangered species in Portugal, the impact of the wind farm on the local population was considered significant, and this led to the development of a site-specific mitigation program (on-site minimization of bird mortality and offset/compensation). The environmental authorities and the wind developer concurred that this was the best solution for reducing kestrel mortality at the wind farm. The mitigation plan included planting native shrubs, enhancing habitat and scrub areas away from the turbines, and promoting extensive livestock grazing away from the turbines to enhance habitat heterogeneity. Monitoring of the kestrel population and carcass surveys continued in order to evaluate the success of the mitigation measures.

4.5 Spain

In Spain, environmental regulators, wind energy companies, and researchers may create agreements that are relevant to AM principles.

Wind farms located in La Janda (Cádiz, southern Spain) provide an example related to finding large numbers of dead birds due to blade collision [26–28]. After several meetings, researchers proposed a method for reducing avian mortality; it consisted of monitoring bird flight in the field, especially the flight of the more affected species such as the Griffon vulture (*Gyps fulvus*). When the wind farm operators detect a dangerous situation, they stop the turbines thought to be likely to harm the birds and restart them after the birds have left the area. Training was provided to operators to ensure accurate detection of collisions during carcass surveys. Daily monitoring for collisions was carried out.

The agreement reached by all parties was as follows: wind energy companies paid for the system; researchers carried out the data analysis and interpretation; and environmental agencies awaited the results before taking further measures. After 2 years, results showed a 50% decrease in mortality and a reduction in energy production of approximately 0.7% per year [27]. Since then, this monitoring method has continued and bird mortality rates continue to decrease.

4.6 Switzerland

Only recently have AM principles for wind energy projects been recognized as a means of improving upon the interpretation of impact assessments and management of bats around wind energy projects [29].

The Gries wind farm in Switzerland, planned to be the highest-altitude wind farm in Europe, has one pilot turbine that has been in operation since 2012 [30]. Three additional turbines are planned, but their potential impacts on migrating bat and bird species are considered to be highly uncertain. As a condition of the construction permit, a curtailment plan will be implemented to mitigate the project's impact on bats. Bat activity will be monitored around the wind energy project for 3 years during spring, summer, and autumn; the resulting data will be assessed on a yearly basis in order to optimize the curtailment algorithm. Any operating adjustments will need approval by an operations commission consisting of stakeholders, the wind farm developers, an independent bat expert, and cantonal (local government) representatives. After 12 years of operation, the commission will reassess the project's operating concept.

4.7 The UK

AM principles are used to manage wind energy and other renewable energy projects in the UK.

In a land-based example, at a 50 MW land-based wind energy project in the UK, developed in moorland habitat over 10 years ago, collision risk models were developed that suggested the facility could pose a risk for hen harriers (*Circus cyaneus*). Monitoring was carried out to determine how to effectively manage the heather moorland habitat to benefit the hen harrier through rotational burning, drain-blocking, etc. The monitoring has found the collision risk is negligible and informs annual decisions about how to best manage the moorland habitat [31].

Offshore, a number of wind farms have been consented in recent years in Scottish waters. Assessments of the potential impacts of wind farms on the marine environment are being used to develop a question-led approach to monitoring that will be able to reduce scientific uncertainties associated with future decision-making. For example, modeling of collision and displacement effects on a range of seabird species, including auks and gulls, is forming the basis of monitoring that is capable of improving confidence in the predictability of seabird behavior in response to operational turbines. Data collection is generally undertaken by consultants acting on behalf of the developers, and the monitoring plans are agreed upon by regional advisory groups with representation from statutory advisors and nongovernmental bodies. Data collection activities serve as a means of validating that the effects associated with the consented wind farm are acceptable, and the results contribute to a double feedback loop that facilitates learning for future offshore wind developments.

4.8 The USA

AM has been applied to wind energy projects in the USA more widely than in European nations. To examine this application, 16 plans governing wind facilities, including bird and bat conservation plans, habitat conservation plans, and biological opinions, were evaluated under the DOI guidelines. More information about the 16 plans can be found in the appendix. Almost all of the plans have a specific focus on federally protected species, such as those protected under the Endangered Species Act, Migratory Bird Treaty Act, and the Bald and Golden Eagle Protection Act [32–34]. In addition, wind energy stakeholders in the USA were interviewed to determine their perceptions of the usefulness and application of AM to wind energy projects. Interviewees included representatives from federal resource management agencies; wind farm developers, owners, and operators; environmental consultants; and nongovernmental organizations.

Prescriptive Adaptive Management Several of the plans reviewed were detailed and contained explicit sections outlining prescribed mitigation actions as a result of specific environmental monitoring results and/or key species mortality events, or triggers, often setting predetermined tiers for monitoring and mitigation thresholds. This approach may provide more certainty for project developers and regulators early in the process.

However, several of the plans reviewed did not contain predetermined mitigation limits or boundaries but did outline a more flexible approach. These plans generally deferred to resource agencies or technical advisory committees to determine mitigation actions and address potential mortalities. The example that most closely followed AM principles was the Grand Prairie Wind Farm plan [35], which contains a tiered structure and outlines a flexible AM approach that relies on a set of triggers or mass mortality events to reengage with regulatory processes. The only offshore wind project reviewed, the Cape Wind project off Massachusetts, also used AM principles to ensure the best available science and technologies are used to monitor and potentially mitigate the project impact on the environment. The Cape Wind project failed to continue toward development.

The need for curtailment and/or compensatory mitigation was the focus of the AM processes described. These plan objectives were intended to inform single-loop learning at the individual project scale rather than emphasizing double-loop learning to inform future wind energy projects.

Role of Mitigation and Compensatory Mitigation in AM Plans The US plans most commonly take the form of bird or bat conservation plans, and they reference a variety of mitigation tools for addressing environmental uncertainty and the unique characteristics of each project. Most of the plans contain pre-established mortality thresholds as mitigation triggers, which may or may not be based on population models. Curtailment of wind power production is the most frequently referenced mitigation response in the plans. All the mitigation examples were developed for land-based wind and may be more difficult to apply to offshore wind.

Compensatory mitigation strategies include power pole retrofitting to prevent eagles from landing on electrical infrastructure to reduce risk of electrocution [36], essential species habitat conservation, and roadside carcass removal to prevent scavenging activities in the vicinity of vehicular traffic. Plans also propose making monetary investments in permanent conservation easements and purchasing critical habitat areas prior to construction [37]. Only a few compensatory mitigation options appear in plans because the number of acceptable options is limited. It is unclear how effective these measures have been.

DOI Guidelines Of the 16 plans examined, 7 outline an AM process that fits the DOI criteria, although precisely fitting the criteria contained in the DOI guidelines proved difficult. The DOI guidelines are very broad, making it problematic to determine whether they are applicable to or relevant for specific projects. Also, several of the AM plans were written for projects that have not yet been constructed and lack the specificity needed to fully understand whether they represent AM as defined by the DOI criteria.

Applying AM – Stakeholders' Perspective The stakeholders who participated in interviews noted that there is considerable variability among AM plans in the USA and that the risk culture among wind farm developers for meeting certain AM requirements is also highly variable.

The interview process confirmed that various definitions of AM are in use. Some stakeholders view AM as a set of tiered, pre-agreed-upon management actions triggered when environmental impacts surpass certain levels, while others argued that true AM should be hypothesis-based. Some stakeholders believed that the flexibility of AM makes it extremely useful, while others claimed AM is a "toolbox without tools," because there is little guidance for developing and implementing AM plans.

Most of the stakeholders believed that the flexibility of AM approaches to environmental uncertainty also creates challenges for financing due to the potential open-endedness of AM plans and associated financial uncertainty. Bounding of mitigation, or setting predetermined mitigation triggers, was mentioned as one way to alleviate some of the uncertainty associated with AM, thereby providing potential investors with a better understanding of the risks, likelihood, and magnitude of the consequences of applying AM.

4.9 Lessons Learned from Case Studies

The case studies presented apply AM principles to management, although many take an approach that is closer to that of the mitigation hierarchy than AM; i.e., they focus on implementing monitoring and mitigation measures to reduce project impacts on a specific bird or bat species. However, a number of examples show how future mitigation measures may be improved at more than the individual project level. The offshore wind farm example from the UK is the closest example to a true AM approach as defined by the US DOI. It illustrates the application of double-loop learning to use stakeholder input, learning by doing, and lessons learned from individual projects to inform future offshore wind project planning and management decisions.

5 Discussion

An evaluation of existing US wind energy conservation plans and the application of AM to wind energy projects worldwide makes it apparent that few examples fully meet the criteria for AM as laid out in the DOI guidelines. Most examples appear to follow the principles of passive, rather than active, AM [4, 9]. It appears that offshore wind energy development is more likely to test active AM.

To systematically decrease scientific uncertainty about interactions between wind energy projects and wildlife, it will be necessary to reconsider the design of monitoring studies at wind energy sites at both the project and planning scales. A necessary first step is to move toward a common definition and application of AM. The overall scale at which AM is applied to wind energy projects and the industry as a whole also should be strongly considered. Studies undertaken at larger spatial scales, or data accumulated from multiple projects, are more likely to correspond closely to a species with a large population range and are therefore likely to be more informative. Sampling efforts at smaller scales are less likely to measure changes in populations, whether or not those changes are due to the presence of the wind farm.

5.1 Consistency and Implementation

The overall approach to and implementation of AM is complicated by the lack of a single definition, as demonstrated by the review of conservation plans from US-based wind energy projects and the application of AM internationally. Within the USA, no federal or state regulations require wind energy projects to use AM, and there is no single definition of AM that is accepted and used by all wind energy projects. However, several natural resource agencies within the USA have recently produced guidelines for AM or rules that invoke the principles of the concept [6, 38, 39]. Similarly, no laws or regulations require AM in any of the other countries evaluated. Existing AM guidelines vary in the degree of prescription of specific actions, the amount of stakeholder involvement and representation in the AM process, and how monitoring and mitigation triggers are developed.

While complying with all national and international regulatory requirements for wind farms, there appear to be opportunities for implementing aspects of AM within existing legal structures and practices. For example, European directives that require preparation of environmental impact assessments for wind energy developments typically follow an approach of "predict-mitigate-implement" for potential wind/wildlife interactions. By encouraging the application of AM principles, while continuing to comply with all applicable regulations, the assessments could be moved toward a "predict-mitigate-implement-monitor-adapt" model that is likely to improve decision-making and ultimately create more effective ecosystem management.

Also, in keeping with the regulatory requirements of many nations, the broad principles of AM require the engagement of stakeholders. Successful AM programs will of necessity involve active information-sharing about scientific questions, potential effects of the development, planned monitoring to decrease uncertainty, and possible mitigation actions if effects are detected. Regulations, such as the US National Environmental Protection Act [40], require stakeholder engagement. Adding the principles of AM could further increase the sense of openness and legitimacy for the process and potentially result in better support of monitoring efforts.

Through this review, we have determined that implementation guidance for AM associated with individual projects, or to benefit future planning for wind energy, should address the following topics:

- The need to develop an AM process from the ground up to meet the specific needs of the location, farm layout, and wildlife in the region, including hypotheses-based questions to guide monitoring activities that will facilitate learning and help to reduce scientific uncertainty.
- The importance of establishing the spatial and temporal scale over which monitoring and data collection should occur, based on the resources of concern and questions to be asked.
- A process for identifying acceptable tolerance levels for monitoring data and potential mitigation triggers.

- A process for engaging stakeholders in decision-making and information-sharing.
- Guidance for integrating or closely connecting AM processes with required environmental assessments (such as Strategic Environmental Assessments in Europe or Programmatic Environmental Impacts Statements in the USA).
- Efforts to determine whether mechanisms can be created that address financial risk, limits, and uncertainties that come about as a result of AM.

More specific guidance could be particularly beneficial for project developers and regulators who choose to use AM but remain unsure of specific implementation practices. Because of the variety of environmental uncertainties and options for policy and site management responses that exist for wind energy, implementation guidelines for wind energy use may be most effective if they are written to allow for a range of monitoring decisions and responses to policies and management of projects.

The overall responsibilities and costs associated with implementing AM and its monitoring activities are typically passed on to the project developer, similar to the polluter pays principle. However, project developers in many European countries may be compensated for part or all of any revenue lost (above an agreed-upon level) due to unforeseen curtailment or mitigation activities. Due to this complexity, which will increase with the scale of implementation, the overall responsibility of who should be required to support and pay for AM may need to be addressed on a case-by-case basis.

5.2 Monitoring to Support AM

Monitoring data needed to successfully support AM processes must be collected using consistent and rigorous methods that are fit for the intended purpose. Most wind energy developers are required to collect baseline data for the area around a proposed wind farm to understand which wildlife populations and habitats might be at risk and to collect data on wildlife interactions with wind farm infrastructure after its construction. In many cases it is difficult to evaluate the effect of wind farm operations on wildlife, particularly at the population level or beyond the scale of any one project. This creates a DRIP situation [19] when monitoring activities are not able to address a hypothesis-driven question at the scale of the population. By initiating an AM process with one or more scientifically valid questions, monitoring programs can yield data that are useful in determining the effects of wind farms on wildlife, provided the data are collected in an appropriate manner. Collection of the data need not require that the overall level of effort be increased but rather that the data collection be thoughtful and aimed at important questions, resulting in a better outcome while not increasing costs to the developer. Such plans should take into account an understanding of species' behavior and relationship to wind turbines. For example, pre-construction monitoring may help predict raptor risk from turbines, but the relationship between pre-construction bat activity levels and post-construction mortality levels has yet to be determined.

5.3 Adaptive Management and the Mitigation Hierarchy

Conservation plans associated with US wind energy projects are usually driven by the need to abide by statutory and regulatory guidance related to numerous environmental laws, resulting in great variability among the plans. Additionally, the relatively recent focus on including AM in monitoring plans has further contributed to variability in these plans. Several of the plans adopt a flexible AM process, while others are more prescriptive and provide an approach to monitoring and mitigation. The latter approach provides developers with certainty about the costs of implementing AM, but unless the study design is fit for its purpose, there is a risk that the outcome will not reduce scientific uncertainty or facilitate learning. In addition, where tiered approaches primarily focus on reducing impacts by monitoring, minimizing, mitigating, and compensating for the effects of wind energy projects, the study design may be less able to detect changes arising from the presence of wind farms and therefore may lack the ability to reduce uncertainty about the mechanisms of the effects. The prescriptive mitigation approach leans toward the overall objective of reducing impacts at each step, which more closely resembles the approach of the mitigation hierarchy than the DOI's definition of AM.

Almost all AM examples reviewed seek to apply AM at an individual project scale and may be relatively ineffective at providing double-loop learning. The ability to provide institutional learning that benefits future planning may be further compromised by the substantial emphasis on reducing impacts at the project scale, as opposed to gathering data to inform a hypothesis. In situations where the mitigation hierarchy is used to minimize potential impacts, such as projects at risk of causing unacceptable impacts on protected or endangered species, an AM process may be inappropriate for helping to inform future management decisions. AM processes may be more useful in determining scientific uncertainty about the efficacy of mitigation or compensation measures rather than scientific uncertainty about the mechanisms of effects causing impacts, in the absence of mitigation or compensation. In practice, as mitigation activities are carried out, an AM approach could be used to evaluate the effectiveness of the mitigation actions, learn from experience, and reduce overall scientific uncertainty by informing more effective mitigation for use in future management decisions. As data are collected, applying AM principles at each level of the mitigation hierarchy (avoidance, minimization, compensation) will allow developers and regulators to maximize learning from their management actions, provided acquired data and knowledge are shared and published broadly.

5.4 *Scale of Implementation*

While AM can be an effective approach for informing mitigation measures and future management decisions and reducing scientific uncertainty for individual projects, the spatial and temporal scale of the monitoring data collected must be appropriate to the questions posed. Applying monitoring results to populations and testing hypotheses with data collected about the impact on individual animals at a single wind energy project is complicated by the large spatial and temporal scales that encompass a species' ecological processes. These scales potentially require large amounts of data to be collected over a long period of time within larger geographic areas, and each wind energy project must be designated as an independent sample rather than as randomly selected samples.

Depending on the project, species of concern, and resources allocated to monitoring, it can be challenging to collect an adequate amount of data to allow for an appropriate statistical power that enables conclusions to be reached with confidence. It is essential that the experiment conducted to examine wind and wildlife interactions be designed at the most relevant spatial and temporal scales, which may be considerably larger than individual wind farms or short time periods after the onset of operation. For example, understanding displacement rates of moorland bird species over time has been most appropriately addressed through meta-analysis of multiple wind farms [41]. There are likely to be persistent challenges to implementing AM at larger spatial and temporal scales that are not derived from scientific concerns but rather from practical limits of data collection and the associated costs.

The temporal and spatial scale at which learning is applied under AM is another important consideration. AM can be applied using double-loop learning or institutional learning (Fig. 1) to apply the lessons learned from other projects to inform future management decisions. Learning outcomes achieved using this approach could have greater overall benefits in terms of protecting wildlife, improving future decision-making, and supporting the further development of the industry.

5.5 *Financial Risks and Mitigation Limits*

As highlighted by many of the US and international stakeholders, AM can lead to unwanted financial risk for a wind energy project. Additional mitigation measures can lead to a decrease in electrical generation in conflict with contracts that provide assurances and set prices for the purchase of power generated by the wind farm, known as power purchase agreements in the USA. In addition, monitoring requirements that are poorly defined initially could become progressively more intensive during plan implementation, thereby leading to more unforeseen project costs. Additional monitoring requirements may include pre-installation assessments and post-installation monitoring that could continue for an undefined period of years.

Collaboratively setting mitigation boundaries or limits has been shown to help address financial uncertainties. Bounding of mitigation involves a negotiation between regulators and project proponents to identify an impact level at which mitigation becomes necessary. Examples of these boundaries can be seen in both US and UK wind energy projects, and they result in a more certain approach to mitigation. Projects that did not follow a prescribed approach tended to rely on resource agencies or technical advisory committees to establish monitoring and mitigation levels and triggers. However, preset boundaries could not be used to remove regulatory authority if significant negative consequences were found after operations began.

Setting mitigation boundaries can also provide regulators with a level of certainty that protection of species and habitats will be maintained and impacts will not be exceeded. However, no mitigation boundaries are absolute; if impacts were found to exceed boundaries, the boundaries could be redrawn or reopened, which may undermine their ability to provide certainty to developers.

6 Conclusions

Wind energy developers and regulators are just beginning to identify the most effective ways in which to balance the strengths and challenges associated with the application of AM. The US DOI guidelines provide an explanation of AM and how the underlying principles may be applied to reducing the scientific uncertainty associated with management of natural resource issues, but consistent application of AM in the context of wind energy regulation does not exist. We make several recommendations for improving the use of AM in wind energy to maximize energy generation while protecting migratory and resident wildlife:

- Specific guidelines or good practices should be created for developing and implementing AM plans for the wind energy industry to address the considerable challenges to applying a concept that does not have a common theoretical foundation agreed upon by practitioners. By building on that foundation and providing specific guidelines for how best to implement AM, increased understanding and support of the approach could be achieved. While it is important to keep AM implementation guidelines at a level suitable for accommodating particular characteristics of individual projects, consistency is needed to allow for comparison between projects and approaches. Only then can key characteristics and attributes that might make the AM process successful be identified, with guidance suggesting that AM plans be as clear as possible. While all wind energy installations must meet regulatory requirements, AM plans should not be perceived as consisting of a loose approach to learning by doing, nor allow opportunities for regulators to exercise additional discretion.

- Well-defined data collection efforts are necessary to answer specific hypothesis-driven questions about interactions between wind energy projects and wildlife. By collecting and aggregating data across wind farms in a consistent manner, the overall scientific uncertainty about these interactions can be reduced and future wind farm development better informed, without significantly increasing financial burdens on developers and wind farm operators of individual projects. Mechanisms of potential harm for wildlife can be best elucidated through strategic research studies that can point toward effective and cost-efficient mitigation measures and potentially decrease overall monitoring needs in the future.
- Establishing an AM process that provides adequate financial certainty is important for project developers and financiers to feel comfortable with implementing an AM process. Establishing tiers, limits, or boundaries for AM activities can help to minimize financial risk. While these mitigation limits and prescribed approaches to AM can be beneficial for minimizing financial risk to a certain extent, they may limit the overall flexibility of the AM process, hampering a project's ability to address future unforeseen issues. Developing appropriate mechanisms and approaches for minimizing the financial risk associated with AM will be critical as more projects begin to rely on AM to address environmental uncertainties.
- The scale at which AM is implemented in the wind energy industry is an important consideration for determining its overall effectiveness. While AM is being applied for some individual projects, challenges remain with measuring change over the spatial and temporal scale of the wildlife resource of concern, which may limit the ability of an individual project to meaningfully reduce scientific uncertainty and facilitate an iterative learning process. To be most effective, the implementation of AM should be considered at a larger spatial and temporal scale than individual projects. The larger spatial and temporal scale of data collection and analysis may consist of a combination of research data collection at the landscape or ecosystem scale with the assistance of public funds and data collection at individual wind farms.

Acknowledgments This chapter is derived from a white paper written under the Working Together to Resolve Environmental Effects of Wind Energy (WREN) initiative (https://tethys.pnnl.gov/publications/assessing-environmental-effects-wren-white-paper-adaptive-management-wind-energy). The WREN international collaborative is a collaboration of 11 nations that seek to identify and resolve conflicts between wind energy development and wildlife protection. The purpose of this chapter is to explore how AM is used in relation to the wind energy industry around the world and to identify ways the process and its implementation may be improved upon. We are grateful to all members of the WREN organization, particularly our co-authors on the white paper – Luke Hanna and Simon Geerlofs of Pacific Northwest National Laboratory, Luke Feinberg of the Bureau of Ocean Energy Management, Jocelyn Brown-Saracino and Patrick Gilman of the US Department of Energy, and Johann Köppel and Lea Bulling of the Berlin Institute of Technology – as well as the many contributors and reviewers who helped improve the white paper.

Appendix: Summary of Wind Development Project Plans with Adaptive Management Components

Project Name	Location	Species of concern	Report name	Motivation for report/regulation referenced	Date
Alta East	Kern County, California	Golden eagle	Conservation Plan for the Avoidance and Minimization of Potential Impacts to Golden Eagles Alta East Wind Project	NEPA, ESA, BGEPA	March 2012
				BLM Instructional Memo (IM) 2010-156, USFWS Draft Eagle Conservation Plan Guidance	
Beech Ridge Wind	Greenbrier and Nicholas Counties, West Virginia	Indiana bat, Virginia big-eared bat	Habitat Conservation Plan (HCP)/Research, Monitoring, and Adaptive Management Plan	ESA – Incidental Take Permit in accordance with a settlement agreement from a lawsuit, NEPA, MBTA, BGEPA	August 2013
Cape Wind	Nantucket Sound, Massachusetts	Roseate tern, piping plover and other avian species, bats	Avian and Bat Monitoring Plan	ESA, MBTA	August 2012
Criterion Wind Farm	Western Maryland	22 rare, threatened, and endangered birds listed in Garrett County, MD, and other eagles	DRAFT Avian Protection Plan	NEPA, BGEPA, MBTA, Maryland Nongame and Endangered Species Conservation Act	March 2012

Criterion Wind Farm	Western Maryland	Indiana bat	Indiana bat HCP-ITP	Incidental Take Permit Application (lawsuit driven)	January 2014
Echanis Wind	Princeton, Oregon	Golden eagle, migratory birds, and bats	Eagle Conservation Plan and Bird and Bat Conservation Strategy	Plan was written for issuance of BLM ROD on the ROW	November 2011
				BLM IM 2010-156	
Grand Prairie Wind Farm	Holt County, Nebraska	Whooping crane, birds, and bats	Bird and Bat Conservation Strategy	MBTA, BGEPA, ESA, Nebraska Regulations	May 2014
Ocotillo Express Wind Energy Facility	Ocotillo, California	Golden eagles	Golden Eagle Conservation Plan for the Ocotillo Wind Energy Facility	Draft Eagle Conservation Plan Guidance	February 2012
Ocotillo Express Wind Energy Facility	Ocotillo, California	Avian and bat species	Avian and Bat Protection Plan for the Ocotillo Wind Energy Facility	USFWS Interim Guidelines for the Development of a Project-Specific Avian and Bat Protection Plan for Wind Energy Facilities (2010)	February 2012
				California Energy Commission Guidelines	
				BLM IM 2010-156	
Shiloh IV Wind Project	Northern California	Bald and golden eagles	Eagle Conservation Plan	BGEPA	June 2014
Spring Valley Wind Farm	Nevada	Eagle, bird, and bat species	Avian and Bat Protection Plan	ESA, MBTA, BGEPA, BLM IM, 2010-156	2010

(continued)

Project Name	Location	Species of concern	Report name	Motivation for report/regulation referenced	Date
Tule Wind Project/Reduced Ridgeline Project	San Diego County, California	Golden eagle, birds, bats	Project-Specific Avian and Bat Protection Plan for the Tule Reduced Ridgeline Wind Project	USFWS Land-based Wind Energy Guidelines	March 2013
Shaffer Mountain	Somerset and Bedford Counties, Pennsylvania	Indiana bat	Biological Opinion; Effects of the Shaffer Mountain Wind Farm on the Indiana bat	Clean Water Act, Endangered Species Act	2011
Mohave County Wind Farm	Arizona	Golden eagles	Eagle Conservation Plan and Bird Conservation Strategy	BLM IM, 2010-156, MBTA, BGEPA, USFWS Land-based Wind Energy Guidelines	2012
Searchlight Wind Energy	Clark County, Nevada	Golden eagle, birds, bats	Bird and Bat Conservation Strategy	ESA, MBTA, BGEPA, Nevada State Codes	2012
				BLM IM, 2010-156, USFWS Land-based Wind Energy Guidelines	
Buckeye Wind	Champaign County, Ohio	Indiana bat	Habitat Conservation Plan	ESA	March 2013

BLM Bureau of Land Management, *BGEPA* Bald and Golden Eagle Protection Act, *ESA* Endangered Species Act, *IM* Instruction Memoranda, *ITP* Incidental Take Permit, *MBTA* Migratory Bird Treaty Act, *ROD* Record of Decision, *ROW* Right of Way, *USFWS* US Fish and Wildlife Service

References

1. Northrup, J.M., Wittemyer, G.: Characterising the impacts of emerging energy development on wildlife, with an eye towards mitigation. Ecol. Lett. **16**, 112–125 (2013). https://doi.org/10.1111/ele.12009
2. May, R., Gill, A.B., Köppel, J., Langston, R.H.W., Reichenbach, M., Scheidat, M., Smallwood, S., Voigt, C.C., Hüppop, O., Portman, M.: Future research directions to reconcile wind turbine–wildlife interactions. In: Köppel, J. (ed.) Proceedings of the Conference on Wind Energy and Wildlife Impacts, Berlin 2015, pp. 255–276. Springer, Cham (2017)
3. Köppel, J., Dahmen, M., Helfrich, J., Schuster, E., Bulling, L.: Cautious but committed: moving toward adaptive planning and operation strategies for renewable energy's wildlife implications. Environ. Manag. **54**(4), 744–755 (2014). https://doi.org/10.1007/s00267-014-0333-8
4. Walters, C.J., Holling, C.S.: Large-scale management experiments and learning by doing. Ecology. **71**, 2060–2068 (1990)
5. Strickland, M.D., Arnett, E.B., Erickson, W.P., Johnson, D.H., Johnson, G.D., Morrison, M.L., Shaffer, J.A., Warren-Hicks, W.: Comprehensive Guide to Studying Wind Energy/Wildlife Interactions. 289 pp. Prepared for the National Wind Coordinating Collaborative, Washington, DC. (2011)
6. Williams, B.K., Szaro, R.C., Shapiro, C.D.: Adaptive Management: The U.S. Department of the Interior Technical Guide. Adaptive Management Working Group, U.S. Department of the Interior, Washington, DC (2009)
7. Williams, B.K., Brown, E.D.: Adaptive Management: The U.S. Department of the Interior Applications Guide. Adaptive Management Working Group, U.S. Department of the Interior, Washington, DC (2012)
8. NRC (National Research Council): Adaptive Management for Water Resources Planning. The National Academies Press, Washington, DC (2004)
9. Murray, C., Marmorek, D.: Adaptive management and ecological restoration. In: Friederici, P. (ed.) Ecological Restoration of Southwestern Ponderosa Pine Forests, A Sourcebook for Research and Application, pp. 417–428. Island Press, Washington, DC (2003)
10. Rogers, K.H., Biggs, H.: Integrating indicators, end points and value systems in the strategic management of the Kruger National Park. Freshw. Biol. **41**, 439–451 (1999)
11. From Resilience to Transformation: The Adaptive Cycle. https://www.google.co.uk/search?q=adaptive+management+double+loop+learning&safe=strict&biw=1280&bih=907&source=lnms&tbm=isch&sa=X&ved=0ahUKEwiy2s-K9bjOAhUHDywKHT9yAg4Q_AUIBigB#safe=strict&tbm=isch&q=adaptive+management&imgrc=oa9fdgdYtk1OzM%3A (2016). Accessed 10 Aug 2016
12. Raffensperger, C., Tickner, J. (eds.): Protecting Public Health and the Environment: Implementing the Precautionary Principle. Island Press, Washington, DC (1999)
13. Kriebel, D., Tickner, J., Epstein, P., Lemons, J., Levins, R., Loechler, E.L., Quinn, M., Rudel, R., Schettler, T., Stoto, M.: The precautionary principle in environmental science. Environ. Health Perspect. **109**(9), 871 (2001)
14. Jakle, A.: Wind Development and Wildlife Mitigation in Wyoming: A Primer. 40 pp. Ruckelshaus Institute of Environment and Natural Resources, Laramie (2012)
15. Kiesecker, J.M., Copeland, H., Pocewicz, A., McKenney, B.: Development by design: blending landscape-level planning with the mitigation hierarchy. Front. Ecol. Environ. **8**(5), 261–266 (2010). https://doi.org/10.1890/090005
16. May, R.: Mitigation for birds. In: Perrow, M. (ed.) Wildlife and Wind Farms: Conflicts and Solutions. Onshore, Solutions; Best Practice, Monitoring and Mitigation, vol. 2, pp. 124–145. Pelagic Publishing, Exeter (2017)
17. Business and Biodiversity Offsets Programme: Mitigation Hierarchy. http://bbop.forest-trends.org/pages/mitigation_hierarchy (2015)

18. MMO (Marine Management Organisation): Review of post-consent offshore wind farm monitoring data associated with licence conditions. A report produced for the Marine Management Organisation, MMO Project No: 1031 (2014). ISBN: 978-1-909452-24-4
19. Ward, R., Loftis, J., McBride, G.: The 'data-rich but information-poor' syndrome in water quality monitoring. Environ. Manag. **10**, 291–297 (1986)
20. German Ministry of the Environment, Agricultural, Nature and Consumer Protection of North Rhine-Westphalia (Translated): Ministerium für Umwelt, Landwirtshaft, Natur- und Verbraucherschutz des Landes Nordrhein-Westfalen (Fassung 10.11.2017). Leitfaden für die Umsetzung von Arten- und Lebensraumschutz bei der Planung und Genehmigung von Windenergieanlagen in Nordrhein-Westfalen. http://artenschutz.naturschutzinformationen.nrw.de/artenschutz/web/babel/media/20171110_nrw%20leitfaden%20wea%20artenhabitatschutz_inkl%20einfuehrungserlass.pdf (2017)
21. Rijkswaterstaat, permit: "BESLUIT inzake aanvraag Wbr-vergunning offshorewindturbinepark 'Q10'" no. WSV/2009-1229, issued 18 December 2009 in Rijswijk, the Netherlands (2009)
22. Wet windenergie op zee [Wind Energy at Sea Act]: As published in "Staatsblad van het Koninkrijk der Nederlanden" [Dutch Government Gazette] on 30 June 2015 (no. 261) in the Netherlands (2015)
23. Bevanger, K., Berntsen, F., Clausen, S., Dahl, E.L., Flagstad, Ø., Follestad, A., Halley, D., Hanssen, F., Johnsen, L., Kvaløy, P., Lund-Hoel, P., May, R., Nygård, T., Pedersen, H.C., Reitan, O., Røskaft, E., Steinheim, Y., Stokke, B., Vang, R.: Pre- and Post-Construction Studies of Conflicts Between Birds and Wind Turbines in Coastal Norway (BirdWind). Report on Findings 2007–2010. NINA Report 620, 152 pp. (2010)
24. Bevanger, K., May, R., Stokke, B.: Landbasert vindkraft. Utfordringer for fugl, flaggermus og rein. NINA Temahefte 66, 72 pp. (2016)
25. Cordeiro, A., Bernardino, J., Mascarenhas, M., Costa, H.: Impacts on Common Kestrels' (*Falco tinnunculus*) populations: the case study of two Portuguese wind farms. Conference on Wind energy and Wildlife impacts. 2–5 May 2011. Trondheim, Norway (2011)
26. Barrios, L., Rodríguez, A.: Behavioural and environmental correlates of soaring-bird mortality at on-shore wind turbines. J. Appl. Ecol. **41**, 72–81 (2004)
27. de Lucas, M., Ferrer, M., Bechard, M.J., Muñoz, A.R.: Griffon vulture mortality at wind farms in southern Spain: Distribution of fatalities and active mitigation measures. Biol. Conserv. **147**, 184–189 (2012)
28. Ferrer, M., de Lucas, M., Janss, G.F.E., Casado, E., Muñoz, A.R., Bechard, M.J., Calabuig, C.P.: Weak relationship between risk assessment studies and recorded mortality in wind farms. J. Appl. Ecol. **49**, 38–46 (2012)
29. Swiss Federal Office of Energy. 2017: "Swiss Federal Office of Energy SFOE – Feed-in Remuneration at Cost." http://www.bfe.admin.ch/themen/00612/02073/index.html?lang=en (2015). Accessed 25 Sept 2017
30. Wellig, S.D., Nusslé, S., Miltner, D., Kohle, O., Glaizot, O., et al.: Mitigating the negative impacts of tall wind turbines on bats: Vertical activity profiles and relationships to wind speed. PLoS One. **13**(3), e0192493 (2018). https://doi.org/10.1371/journal.pone.0192493
31. Forrest, J., Robinson, C., Hommel, C., Craib, J.: Flight activity & breeding success of Hen Harrier at Paul's Hill Wind Farm in North East Scotland. Poster presented at the Conference on Wind Energy and Wildlife Impacts, Trondheim, Norway, 2–5 May 2011,
32. Endangered Species Act. 1973. 16 USC § 1531
33. Migratory Bird Treaty Act. 1916. 16 USC § 1531
34. Bald and Golden Eagle Protection Act. 1972. 16 USC § 668
35. Stantec (Stantec Consulting Services, Inc.).: Final Buckeye Wind Power Project Habitat Conservation Plan. Edmonton, Canada (2013)
36. Cole, S., Dahl, E.L.: Compensating white-tailed eagle mortality at the Smøla wind-power plant using electrocution prevention measures. Wildlife Soc. B. **37**, 84–93 (2013). https://doi.org/10.1002/wsb.263

37. Criterion Power Partners, LLC.: Indiana Bat Habitat Conservation Plan for the Criterion Wind Project, Garrett County, Maryland. Oakland, Maryland. http://www.fws.gov/chesapeakebay/endsppweb/Criterion%20docs/FINAL%20Criterion%20HCP.pdf (2014)
38. BLM (Bureau of Land Management).: "Instruction Memorandum No. 2010-156 – Bald and Golden Eagle Protection Act – Golden Eagle National Environmental Policy Act and Avian Protection Plan Guidance for Renewable Energy." July 9. http://www.blm.gov/wo/st/en/info/regulations/Instruction_Memos_and_Bulletins/national_instruction/2010/IM_2010-156.html (2010)
39. USFWS (US Fish and Wildlife Service).: Eagle Conservation Plan Guidance Module 1 – Land Based Wind Energy Version 2. (2013). http://www.fws.gov/windenergy/PDF/Eagle%20Conservation%20Plan%20Guidance-Module%201.pdf
40. National Environmental Protection Act. 1969. 42 USC §4321 et seq
41. Pearce-Higgins, J.W., Stephen, L., Douse, A., Langston, R.H.W.: Greater impacts of wind farms on bird populations during construction than subsequent operation: results of a multi-site and multi-species analysis. J. Appl. Ecol. **49**, 386–394 (2012). https://doi.org/10.1111/j.1365-2664.2012.02110.x

Wildlife Mortality at Wind Facilities: How We Know What We Know How We Might Mislead Ourselves, and How We Set Our Future Course

Manuela Huso

Abstract To accurately estimate per turbine – or per megawatt – annual wildlife mortality at wind facilities, the raw counts of carcasses found must be adjusted for four major sources of imperfect detection: (1) fatalities that occur outside the monitoring period; (2) carcasses that land outside the monitored area; (3) carcasses that are removed by scavengers or deteriorate beyond recognition prior to detection; and (4) carcasses that remain undiscovered by searchers even when present. To accurately estimate regional or national annual wildlife mortality, data must come from a representative (or appropriately weighted) sample of facilities for which estimates of mortality account for all sources of imperfect detection. I argue that the currently available data in the United States and much of the world do not represent the impacts of wind power on wildlife because not all facilities conduct monitoring studies, not all study results are publicly available, and few studies adequately account for imperfect detection. I present examples illustrating the limitations of our current data and pitfalls of interpreting data without accurately adjusting for detection bias. I close by proposing a solution through a simplified monitoring process that can be applied at every facility as part of normal operations. Application of an unbiased estimator that accounts for all sources of imperfect detection would assure comparability of mortality estimates. Public access to reported estimates would achieve representation. With these data we could develop a clearer understanding of how wind power is affecting wildlife throughout the world and inform our efforts to address it.

Keywords Representative · Spatial distribution · Road and pad searches · R&P · Detection probability

M. Huso (✉)
US Geological Survey, Corvallis, OR, USA
e-mail: mhuso@usgs.gov

R. Bispo et al. (eds.), *Wind Energy and Wildlife Impacts*,
https://doi.org/10.1007/978-3-030-05520-2_2

1 Introduction

In trying to understand the impact of wind power development on wildlife, what might we want to know? We might start with: Is there a problem? If we knew how many birds and bats are killed in the United States or Europe or any other region, we could evaluate that number relative to other sources of impact on wildlife and as a society decide whether the impact is one whose cost we are willing to bear in order to reap the benefits of renewable energy. But it is not strictly a question of how many birds or bats are killed. For example, different species have different perceived value, and the death of a European starling (*Sturnus vulgaris*) in the United States does not have the same societal relevance as the death of a golden eagle (*Aquila chrysaetos*). We also might want to ask whether certain species are differentially affected by wind power development and whether their populations can be expected to persist and thrive despite the toll taken by wind turbines. Further, to minimize impacts we would like to be able to predict where or when high mortality will occur in order to avoid development in those areas or operations during those times.

The most important question to answer before any of these can be answered is: Do we have the data with which to address these questions, and if not, how can we assure that we will? To accurately estimate fatality at a wind facility, the raw counts of carcasses found must be adjusted for four major sources of imperfect detection of all carcasses killed annually by wind turbines: (1) fatalities that occur outside the monitoring period; (2) carcasses that land outside the monitored area; (3) carcasses that are removed by scavengers or deteriorate beyond recognition prior to detection; and (4) carcasses that remain undiscovered by searchers even when present. The latter two factors, removal by predators and searcher inefficiency, have long been recognized as important [1, 2]. The second factor was initially largely ignored but has recently come more to the forefront [3], while the first, temporal coverage of the monitoring period, is still largely ignored. Any analysis that seeks to summarize cumulative effects of wind power or compare magnitude of mortality among turbines, sites, or regions will need to be conducted using data that represent a common metric. If any of these factors is ignored or incorrectly accounted for, the numbers derived will not be comparable to those derived from studies in which full accounting is appropriately applied. For example, estimated bat mortality from a study conducted in Northeastern Germany between August and October cannot be meaningfully compared to one in Spain conducted from March through November without accounting for the very different monitoring period. Although in these areas the majority of fatalities appear to occur during the late summer and early fall, it is by no means all. When compared or used in cumulative analyses, mortality estimates should be standardized to reflect comparable periods.

Current practices for monitoring and reporting differ greatly among countries and often among states or provinces within countries. Some require monitoring and public access to data from all facilities. Others do not require monitoring at all. Some require monitoring, but not public release of the results. This variation in requirements, by no means unique to the United States, results in high degree of variation in availability of data. The data that do end up publicly available cannot be considered representative of the full range of facilities and hence are not representative of the total impact of wind power on wildlife. In addition, approaches used to account for imperfect detection vary greatly among studies: some completely ignore differences in detectability and report simple carcass counts; some account for some components of non-detection, e.g., searcher efficiency or scavenging, while ignoring others, e.g., fraction of fatalities in the searched area; yet others are well-designed studies that appropriately measure and account for all sources of imperfect detection. Combining information from these various studies without accounting for the large differences in quality and meaning of the reported values will lead to misunderstanding of the actual impact of wind power development on wildlife. Like the proverbial blind men and the elephant, our arguments will be based on an inaccurate picture of the whole. Unless we appropriately account for all the ways in which our monitoring process might miss some fatalities and unless we have a representative sample of the wind power facilities about which we are drawing inference, we will develop a very biased picture of the impact of wind power development on wildlife on a continental regional and even local level.

2 Examples

In this section I will briefly review several examples illustrating the limitations of our current data and the pitfalls of interpreting data without accurately adjusting for detection bias. The examples all come out of the published literature, both peer-reviewed and publicly available reports, with one exception indicated below. They are all from North America, not because the limitations do not exist elsewhere but because at a European conference I try to bring an outside perspective and a focus on a literature with which conference attendees may not be so familiar. The point is not to criticize the authors but to point out places where mistakes are commonly made and how these mistakes can lead to misinformation and misguided perception of the impacts of wind power development on wildlife.

2.1 Representative Data

I will start with the question “Is there a problem?” Several authors have tried to answer this question for the United States. All are limited by the available data. I will focus on a couple of these studies for two reasons – one to highlight the difficulties in using currently available data to understand large-scale patterns and the other to emphasize our individual responsibilities to think carefully about the validity of our analyses and the impacts of our statements in the scientific literature. One study recently cited a “large number” of bats (600,000) being killed annually at wind facilities in the United States [4]. Is this something we should be concerned about? How does it compare to the overall population of bats in the United States? More importantly, is mortality of "all bats" the most appropriate metric? What is the impact to individual species, facing different degrees of threat to their populations? 600,000 of a species with six billion individuals surely have different import than 600,000 of a species with only six million.

Before we can interpret this estimate, we should determine if the methods used to calculate this number could be expected to result in an accurate estimate [5]. To arrive at such an estimate, a tried-and-true statistical approach is to take a random sample of all facilities or all turbines or all megawatt (MW) capacity in the region for which the estimate is needed to obtain a representative sample. At each of these selected units, we would count the number of carcasses observed, estimate the probability that a carcass made it through the gantlet of reasons why it might be missed, appropriately adjust the count for this detection bias, and estimate mortality. If the estimator used is unbiased, then estimates from each of these units can be combined with others to estimate a regional total (with associated confidence intervals, of course).

In this study [4] the author explicitly assumed his data were from a representative sample of wind energy facilities in the United States yet acknowledged that several regions have little representation in the available data. He used 22 data points in his analysis, i.e., 22 estimates of fatality reported from throughout the United States (Fig. 1), to represent the 60-GW capacity in the United States at the time. It was primarily a convenience sample compiled for use in another publication by other authors for another purpose [6], to which he added three arbitrary points. Printed on a map of the United States, the points seem fairly well scattered geographically and hence seemingly well representative (Fig. 1a). However, for 22 data points to be representative of the total effect of wind power on bat fatality, we would expect to have about 1 point for every 5% of the total build out either in terms of turbines or MW capacity or power produced. I will focus on MW capacity as the number of birds or bats per MW capacity seems to be a societally relevant measure in the current literature.

In the West, in 2012, the reference year used in this study, about 19% of wind power generation capacity was in Washington, Oregon, and California (Fig. 1b), so we might expect around four data points from this region, and indeed five data points (23% of the sample) came from that region. The central states comprised about 25% of total wind capacity in 2012, but data came from only two locations (one of these was entered twice in the data set, once as raw counts, once with adjustments for searcher efficiency and carcass persistence, but no adjustment for the 20 m search radius nor the 2-month monitoring period). Again, in the Midwest only two (one measured three times) comprised 9% of the sample, for 23% of the capacity. Texas alone had 20% of the total capacity of US wind power at that time, yet only one study from Texas (5% of the sample) was included. The remaining 11 observations (50% of the total sample) were from the Northeast, Wisconsin, and Tennessee although those states comprised only 9% of the total capacity. And those 11 data points came from only 6 locations: One location in West Virginia, with the second highest fatality estimates in the data set, was included twice. And a single site in Tennessee, with the

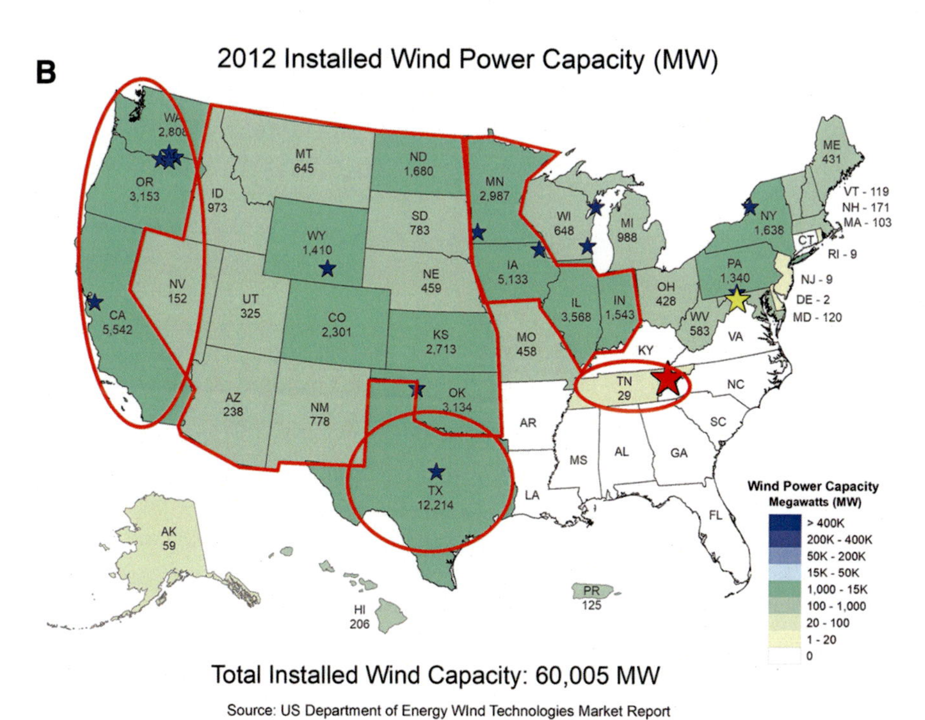

Fig. 1 (**a**) Location of sites used to estimate total annual US bat mortality. (**b**) MW capacity of each state and region. A single site represented by the red star in Tennessee, a state with 0.05% of total US capacity, comprised 3.6% of the sample data. A single site represented by the yellow star in West Virginia, a state with 0.1% of total US capacity, comprised 9% of the sample data

highest fatality estimates in this data set, was included three times. Tennessee, with 5/100 of a percent of the total MW capacity of the United States had three times the weight of Texas that had 20% of the total MW capacity of the United States.

There is clearly nothing to suggest that these 22 values from 16 facilities are a representative sample of the facilities in the United States nor have all the values reported been standardized to represent the estimated mortality per MW capacity at the site over a fixed period. In addition, selection from the peer-reviewed literature may itself be an inherently biased sample [7, 8]. So, we have no reason to believe that an estimate of total bat mortality derived from this sample represents the fatality impacts of wind power in the United States. Unfortunately, it is also not clear that any sample of the currently available data could be used to derive an accurate estimate. Monitoring requirements differs greatly across political boundaries, as do reporting requirements. Not all facilities are monitored, data from monitored facilities are not all publicly available, and the likelihood of publicly reporting results might be influenced by the measured mortality itself. Estimates might be more likely to be voluntarily reported if they are perceived to be low. On the other hand, studies might be more likely to be published in the peer-reviewed literature if they are perceived to be unusually high. In addition, attempts to correct for detection bias among monitored facilities vary greatly, ranging from none to very complete. As John Tukey [9] so astutely noted "The combination of some data and an aching desire for an answer does not ensure that a reasonable answer can be extracted from a given body of data." This is the problem we are squarely facing with our currently available mortality data at wind power facilities in the United States and in most parts of the world.

2.2 Comparing or Interpreting Raw Counts

A question that is at least equally as important as how many total individuals of a group are being killed continent-wide is whether there are some species or perhaps species groups that are differentially affected by wind. Many people have tried to address this by producing pie charts (or tabular summaries) that divide the observed fatalities into species groups and assign each a proportion of the total. From summaries like this, relative impact to different species groups is inferred. Unfortunately, these charts are often constructed using raw counts, ignoring differential detection rates among individuals as well as among groups. If detection rates can reasonably be assumed constant across groups as might be reasonable for bats in some regions of the world, then tables or charts of percent composition can be meaningful [10]. However, if detection rates are substantially different among individuals, e.g., large vs small bird, percent composition summaries can be very misleading.

For example, because 3110 of 4975 birds reported in the studies summarized by Erickson et al. [11] were small passerines, they estimated that passerines comprised 62.5% of fatalities (Fig. 2a). From this type of graph and reported values, one is

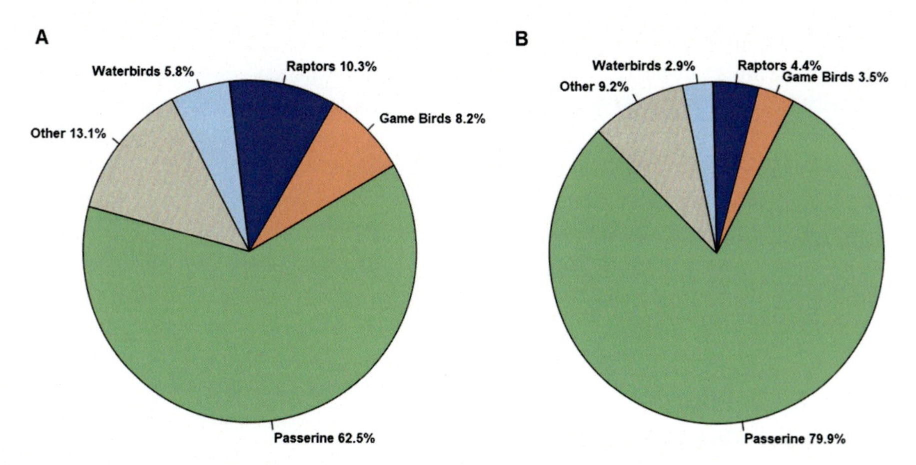

Fig. 2 (**a**) Percent of observed carcasses broken down by species groups. (From Erickson et al. [11], Table 3). (**b**) The same data, adjusted for assumed detection probabilities

tempted to infer that there were six times more passerines killed than raptors. This, however, ignores a critical component in appropriate representation of the data – the very unequal detection rates among carcasses even within a single site or under a single turbine. Generally larger carcasses persist longer and are easier to find than smaller ones. Generally, raptors are larger than passerines.

Reporting raw numbers as percent of carcasses found when detection is not constant is meaningless unless perhaps you are planning the freezer space you will need for storing carcasses. When the raw counts are adjusted for the variable probability of detection among carcasses, the picture can change radically. If we make a very coarse assumption simply for illustrative purposes that detection probabilities for small, medium, and large birds were 0.25, 0.5, and 0.75 respectively, the proportions change significantly (Fig. 2b). Proportional representation of groups comprised primarily of larger birds decreases, while that comprised of smaller birds increases. After accounting for differential detection probabilities, in this illustrative case, a more reasonable estimate would be that closer to 18 times more passerines were killed than raptors, not 6 times. The authors further used this proportion calculated using only raw numbers to adjust their own estimate of the "number of small birds killed by turbines (133,993 and 229,765; Table 5) with the proportion of all fatalities that were passerines (62.5%), [to calculate] that about 214,000 to 368,000 turbine-related deaths occur each year for all birds." Had they accounted for differential detection probabilities in their calculation of the proportion comprised of small birds, their all-bird estimates would likely have been lower.

Using proportional composition of discovered carcasses can lead to absurd estimates. In one Canadian study [12], the estimated total national mortality was split into species-specific mortality based on the percent composition of each species in the raw counts of discovered carcasses. The unfortunate consequence of this approach is that species with similar counts will be predicted to have similar mortality, regardless of their general ease or difficulty of detection. For example, because purple martins (*Progne subis*) and mallards (*Anas platyrhynchos*) both numbered 20 in the observed species list [12, Table 2], national total mortality of the two was predicted to be equal. It is, of course, possible that the two are approximately equal, but this would be a serendipitous coincidence. It is much more likely that detection probability for mallards was generally greater than that of purple martins and that the number of purple martins missed during monitoring for every one found would exceed the number of mallards missed for an equivalent number found. Basing mortality estimates on species composition in lists of raw counts can be misleading.

2.3 Averaging Detection Probability Across Classes

Cognizant that the probability of detection is something that cannot be ignored and that some reports failed to include estimates of detection probability, the solution of some authors has been to "adjust fatality estimates ... [using] national averages from hundreds of carcass placement trials intended to characterize scavenger removal and searcher detection rates" by applying the average values to the reported raw counts from all facilities [13]. This same line of reasoning is sometimes applied at the study level as well, where reports list "average" searcher efficiency and/or carcass persistence, calculated over several visibility classes or seasons [14]. If hypothesis tests or model selection processes provide no evidence that searcher efficiency differs among classes within a study and sample sizes are large enough to detect meaningful differences, then calculation of an average may be warranted. However, when estimated detection rates differ substantially (say more than by 0.2), the average across classes has little meaning, and applying it to all carcasses regardless of class can lead to inaccurate estimates. The average detection probability is based only on the number and proportion of trial carcasses placed in each class and could vary markedly simply because more or fewer carcasses were available for trials.

I used an example derived from a published report [14] to illustrate Table 1.

In this example, searcher efficiency (SE) appears fairly constant among seasons within the grassland habitat, and an F-test of differing means shows no evidence of such. Greatly varying arrival rates across season would have little effect, and averaging across these values of searcher efficiency would be reasonable. However, searcher efficiency varies greatly among seasons in the agricultural habitat. For an average searcher efficiency across seasons to be accurate, it would have to be weighted relative to carcass arrivals in each season. The problem is that we do not know the relative arrival rates, so we cannot calculate a meaningful weighted average.

Table 1 Searcher efficiency trial results from four seasons in two habitat classes

Grassland	Placed	Found	SE
Winter	11	5	45%
Spring	7	3	43%
Summer	13	6	46%
Fall	19	9	47%
Subtotal	85	36	42%
Agriculture	**Placed**	**Found**	**SE**
Winter	14	4	28%
Spring	7	6	86%
Summer	9	1	11%
Fall	5	2	40%
Subtotal	35	13	37%

Given the current allocation of trial carcasses in each season, the average detection probability in the agricultural habitat, ignoring season, is calculated as $(4 + 6 + 1 + 2)/35 = 37\%$. But what if the effort had been reversed, with 14 placed in spring because that is when highest mortality was expected and only 7 in winter? The detection probability in each class does not change, but the overall average does. With a searcher efficiency of 28%, one would expect to observe around 2 out of 7, on average in the winter, and with a searcher efficiency of 86%, one would expect to observe around 12 out of 14, on average, in the spring. The average detection probability, ignoring class, would be calculated as $(2 + 12 + 1 + 2)/35 = 49\%$, far different from the original calculation simply because trail carcasses were allocated differently. Most importantly, we have no reason to believe that either average represents the average detection rate among all carcasses that arrive at a site because we do not know what fraction will arrive in each class. Unless the proportions of trial carcasses placed in each class also parallel the arrival patterns of carcasses to these classes, an average detection probability across classes will not be accurate, nor will mortality estimated from it.

Although applying this process at the national scale might seem an attractive way to deal with the varying efforts to account for imperfect detection, it is equally problematic. One author attempting to estimate national annual wind turbine-caused mortality used a national average of hundreds of carcass placement trials to adjust all reported counts [13]. Average detection rates when trial sizes are not proportional to arrival have no meaning and when applied on a national or regional scale can result in absurd estimates.

2.4 *Carcass Fall Patterns*

Plot area and configuration vary greatly among studies. For example, the searched areas in the 22 studies in the first example varied from 20 m radius to 100 m radius around turbines, yet few of the studies accounted for carcasses falling beyond plot boundaries in their estimates. Often, even within plot boundaries, there are areas that are inaccessible or unsearchable. Because the density of carcasses changes with distance from the turbine, the exact location and configuration of the unsearched area are needed to accurately estimate the fraction of carcasses expected to land within it [3].

Unfortunately, too often, distance distributions of found carcasses are reported without accounting for the fraction of area in any given distance class that was actually searched. For example, a study at a wind facility in the Midwestern United States, located within an agricultural area, used a unique mowing pattern to provide some easily accessible paths within the dense corn [15]. At 10% of the turbines, they mowed and searched the entire area within a 160 × 160 m plot centered on the turbine. At 90% of the turbines, they searched a 160 × 10 m strip along the major axis (including the access road) as well as five randomly selected swaths perpendicular to this axis (Fig. 3a). This resulted in varying proportion of area within each distance band that was actually searched (Fig. 3b).

They reported the number of carcasses found within each distance band (Fig. 4a green) and the average distance at which carcasses were found, without accounting for the varying proportion of area searched within the band (Fig. 3b). While 100% of the area within 10 m was searched, only 25–33% of the area in bands beyond that were searched and even less beyond 80 m. A very different distribution emerges when the proportion of area searched within each band is taken into account (Fig. 4a grey). Although 42% of carcasses were found within 20 m of the turbine, a more accurate estimate that accounts for area searched would be 26% (Fig. 2b). A single carcass was found over 100 m from the turbine, and with only 2.5% of

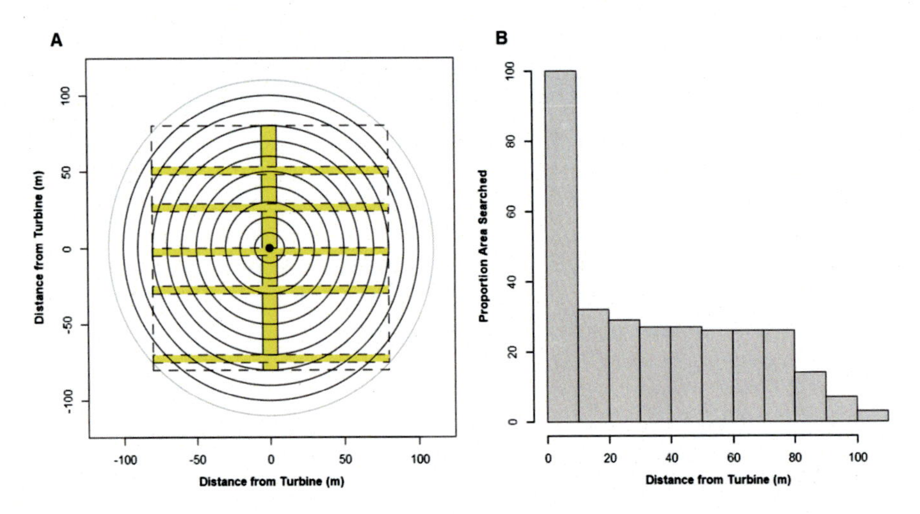

Fig. 3 (**a**) Configuration of search transects cut into agricultural field. (**b**) Proportion of area searched within each 10 m distance band

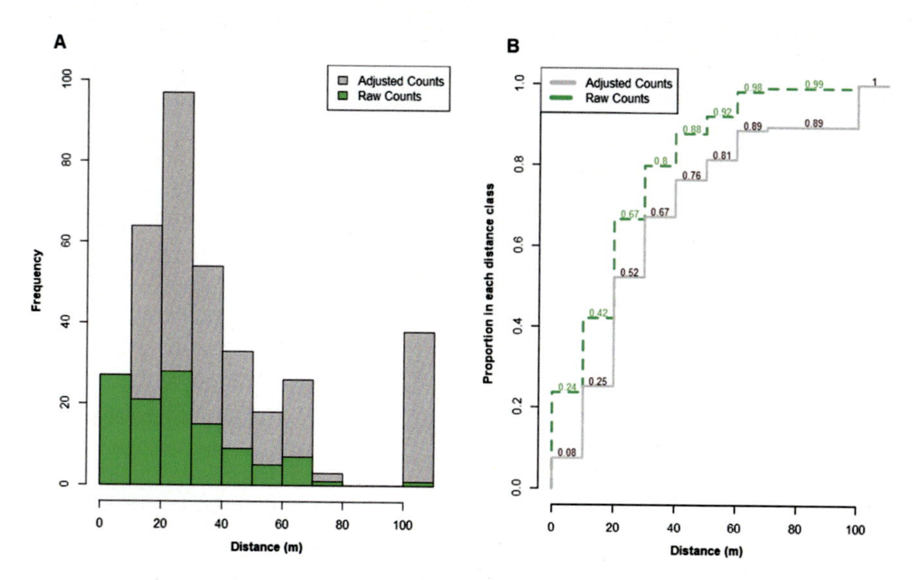

Fig. 4 (**a**) Histogram of raw counts (green) and counts adjusted for proportion of area searched (grey) within 160 × 160 m plot. (**b**) Cumulative proportion of total carcass count (green) and count adjusted for proportion of area searched (black) within each distance class

the area in that band searched, it represents a large number of carcasses that were potentially missed. While this might be an unfortunate chance event, it also might be an indication that some, perhaps nontrivial, proportion of carcasses can land beyond 90 m.

The implications are that smaller plots might comprise far less of the total fatality than we currently assume. This does not necessarily mean we should search larger plots. It only means that we need to accurately account for the spatial distribution of carcasses to be able to account for the fraction falling outside the limits of our searched area. This is a critical component in accurately estimating fatality at any facility.

2.5 *Summary of Issues*

In this section I have highlighted some cautionary tales to consider when analyzing and reporting results from post-construction fatality monitoring studies. They all stem from two sources: the available data do not represent the true impact of wind power development on wildlife; or the analysis inadequately accounted for all sources of imperfect detection. In the following section I propose a simple solution to both problems.

3 Proposed Solution

“It is not the critic who counts; not the [one] who points out how [another] stumbles ... The credit belongs to the [one] who is actually in the arena, ... who errs, who comes short again and again, because there is no effort without error and shortcoming” [16]. I bring up these examples not to focus on the errors of a few – we all make mistakes – but to highlight the limitations of the data we currently have available to us and the need, in almost every case, to analyze fatality estimates, i.e., counts properly adjusted for imperfect detection, not the raw counts themselves. What we think we know so far is likely a biased picture of what is actually happening. We must be constantly vigilant to resist the bias that comes from only a myopic view of the world and acutely aware of the impacts our published work can have on the perceptions of our community.

I return to the original questions – is there a problem? Where and when? For which species? Can we answer these questions? My response to the last is “No, not with the data we currently have available... but I believe we could.” We could if we had an approach that resulted in a representative sample of the facilities in all regions and continues to be representative as wind power increases and adopts constantly improving technologies (with their potentially differing levels of impacts) and expands into areas as yet undeveloped for wind. We could if we had an approach that could result in unbiased estimates of fatality from data collected at this representative sample of facilities. We could if we took advantage of many of the tools that have been developed in our field over the last few years – better models of the spatial distribution of carcasses to account for carcasses falling outside the searched area [3]; better models of persistence beyond assuming constant removal [17]; better models of searcher efficiency beyond assuming constant detection with each search [18]; and better integration of these to form estimates of probability of detection that are accurate and essentially carcass specific [19].

Consider a monitoring program that called for carcass surveys (and associated detection trials) of every facility on every turbine for some X (not necessarily continuous) years of operation but with searches conducted only on areas with high probability of detection, e.g., roads and pads (R&P). If data collected under such a program were publicly available, the issue of representation would be moot. A R&P search protocol can be easily standardized leading to more predictable costs. Overall costs would likely be reduced relative to current approaches because physical area searched would be smaller or more easily accessed and trial effort would be reduced because searcher efficiency and carcass persistence would need to be estimated for only a single visibility class. Limitations in searchable area due to crops, cliffs, or cobras (CCC) result in highly variable efforts across facilities, whereas R&P searches at all turbines would be much more consistent and predictable.

As shown earlier, the maximum distance a carcass might fall from a turbine is likely greater than we currently believe, so complete coverage of plots would be extremely expensive, with diminishing returns for effort as distance from turbine increases. Searches along roads, on the other hand, can easily begin far from the turbine (the area must be traversed anyway as one approaches the turbine) at little additional cost yet provides data critical for developing models of carcass spatial distributions needed to estimate the fraction of carcasses within the R&P searched area. While limiting searches to only R&P might reduce precision of fatality estimates at any given facility, this statistical cost would be made up by a great increase in coverage and hence representation of all facilities. Newly developed estimation tools will provide unbiased estimates of mortality from R&P [19], achieving comparability of information across facilities and regions. Concerns about plot size, properly accounting for all visibility classes, or other factors influencing detection will be removed. With unbiased estimates of fatality and all facilities represented, we will be able to develop realistic estimates of the actual impact of wind power development. Spatial analysis can identify centers of concern but also identify areas of minimal impact.

This is a proposal to address baseline monitoring, not an optimal approach if particular research questions are of interest or studies require a generally high probability of detection to address evidence of absence of species of concern. But it would resolve many of our current inferential issues if applied regularly and consistently across all facilities throughout their period of operation. The details of this proposal are not yet worked out, but I hope I have generated some food for thought and initiated a healthy discussion about where we might take our understanding of wind-wildlife interactions over the next 10 or more years. We all have different world views and different tolerances for the level of impact we feel is acceptable, but the argument should be informed by accurate and meaningful data. Right now, for us to really understand the impacts of wind power and to be able to act to reduce them, we need to be extremely careful in how we evaluate our current data. "No ingenuity of statistical analysis can force a badly designed experiment to yield evidence on points to which insufficient thought was given in the designing, whereas even an imperfect or inefficient analysis of a well-designed experiment may give conclusions that are enlightening and substantially correct" [20]. One solution is to design our post-construction monitoring process collectively and start collecting data, at relatively low cost, that actually allow meaningful comparison across regions and across facilities. If these data were reported and made publicly available from every project, representation would not be an issue. This is not a new request; several authors have called for transparent reporting of results [21–23], but to date, collision data remain largely inaccessible. And although the variance on our estimates at individual facilities might be larger than with more intense surveys, the coverage of our sampling and the unbiasedness of our estimator would allow us to see pattern – and from there start to develop a much clearer understanding of how wind power development is affecting wildlife and what we can do about it.

Acknowledgments I thank the organizing committee of the Conference on Wind Wildlife for inviting me to give this talk and supporting my travel. I thank Fränzi Korner-Nievergelt, Mona Khalil, and an anonymous reviewer for helpful suggestions to improve the manuscript. I thank Dan Dalthorp, Guillaume Marchais, Cris Hein, Michael Schirmacher, and Jerry Roppe, for many fruitful discussions regarding the feasibility of road and pad searches, but do not wish to imply any endorsement on their part of the ideas I present here. Funding for this research was provided by the Ecosystems Mission Area Wildlife Program of the US Geological Survey (USGS). Any use of trade, firm, or product names is for descriptive purposes only and does not imply endorsement by the US Government.

References

1. Rogers, S.E., Cornaby, B.W., Rodman, C.W., Sticksel, P.R., Tolle, D.A.: Environmental Studies Related to the Operation of Wind Energy Conversion Systems. U.S. Department of Energy, Columbus (1977)
2. Orloff, S., Flannery, A.: Wind Turbine Effects on Avian Activity, Habitat Use, and Mortality in Altamont Pass and Solano County Wind Resource Areas, 1989–1991 In: vol. 23 May 2008. California Energy Commission, Sacramento (1992)
3. Huso, M.M.P., Dalthorp, D.H.: Accounting for unsearched areas in estimating wind turbine-caused fatality. J. Wildl. Manag. **78**(2), 347–358 (2014). https://doi.org/10.1002/jwmg.663
4. Hayes, M.A.: Bats killed in large numbers at United States wind energy facilities. Bioscience. **63**(12), 975–979 (2013). https://doi.org/10.1525/bio.2013.63.12.10
5. Huso, M.M.P., Dalthorp, D.H.: A comment on "Bats Killed in Large Numbers at United States Wind Energy Facilities". Bioscience. **64**(6), 546–547 (2014). https://doi.org/10.1093/biosci/biu056
6. Arnett, E.B., Brown, K.W., Erickson, W.P., Feidler, J.K., Hamilton, B.L., Henry, T.H., Jain, A., Johnson, G.D., Kerns, J., Koford, R.R., Nicholson, C.P., O'Connell, T.J., Piorkowski, M.D., Tankersley Jr., R.D.: Patterns of bat fatalities at wind energy facilities in North America. J. Wildl. Manag. **72**, 61–78 (2008)
7. Ioannidis, J.P.A.: Why most published research findings are false. PLoS Med. **2**(8), e124 (2005). https://doi.org/10.1371/journal.pmed.0020124
8. Scargle, J.D.: Publication bias: the "file drawer" problem in scientific inference. J. Sci. Explor. **14**(1), 91–106 (2000)
9. Tukey, J.W.: Sunset Salvo. Am. Stat. **40**(1), 72–76 (1986). https://doi.org/10.2307/2683137
10. Johnson, G.D., Perlik, M.K., Erickson, W.P., Strickland, M.D.: Bat activity, composition and collision mortality at a large wind plant in Minnesota. Wildl. Soc. Bull. **32**(4), 1278–1288 (2004)
11. Erickson, W.P., Wolfe, M., Bay, K.J., Johnson, D.H., Gehring, J.L.: A comprehensive analysis of small-paserine fatalities from collision with turbines at wind energy facilities. PLoS One. **9**(9), 1–18 (2014). https://doi.org/10.1371/journal.pone.0107491
12. Zimmerling, J.R., Pomeroy, A.C., d'Entremont, M.V., Francis, C.M.: Canadian estimate of bird mortality due to collisions and direct habitat loss associated with wind turbine developments. Avian Conserv. Ecol. **8**(2), 10 (2013). https://doi.org/10.5751/ACE-00609-080210
13. Smallwood, K.S.: Comparing bird and bat fatality-rate estimates among North American wind-energy projects. Wildl. Soc. Bull. **37**(1), 19–33 (2013)
14. Erickson, W.P., Kronner, K., Gritski, B.: Stateline Wind Project Wildlife Monitoring Final Report July 2001–December 2003. In: WEST, Inc., p. 98. FPL Energy, Cheyenne (2004)
15. Grodsky, S.M., Drake, D.: Assessing Bird and Bat Mortality at the Forward Energy Center in Southeastern Wisconsin. In: vol. PSC Ref#: 152052. Forward Energy LLC, Chicago (2011)
16. Roosevelt, T.: Citizenship in a Republic. Paper presented at the delivered at the Sorbonne, Paris, France (1910)
17. Bispo, R., Bernardino, J., Marques, T., Pestana, D.: Modeling carcass removal time for avian mortality assessment in wind farms using survival analysis. Environ. Ecol. Stat. **20**(1), 147–165 (2013). https://doi.org/10.1007/s10651-012-0212-5

18. Wolpert, R.: Appendix B: a partially periodic equation for estimating avian mortality rates. In: Warren-Hicks, W., Newman, J., Wolpert, R., Karas, B., Tran, L. (eds.) Improving Methods for Estimating Fatality of Birds and Bats at Wind Energy Facilities, CEC-500-2012-086, pp. A1–A20. California Wind Energy Association, Berkely (2013)
19. Hein, C., Huso, M.M.P.: Challenges and opportunities of accurately and precisely estimating fatality of birds and bats at wind. Paper presented at the Ecological Society of America, Portland, OR, USA, Aug 6–11 (2017)
20. Finney, D.J.: Statistical Method in Biological Assay. Charles Griffin & Co. Ltd, London (1952)
21. Kunz, T.H., Arnett, E.B., Erickson, W.P., Hoar, A.R., Johnson, G.D., Larkin, R.P., Strickland, M.D., Thresher, R.W., Tuttle, M.D.: Ecological impacts of wind energy development on bats: questions, research needs, and hypotheses. Front. Ecol. **5**(6), 315–324 (2007)
22. Loss, S.R., Will, T., Marra, P.P.: Estimates of bird collision mortality at wind facilities in the contiguous United States. Biol. Conserv. **168**, 201–209 (2013)
23. Piorkowski, M.D., Farnsworth, A.J., Fry, M., Rohrbaugh, R.W., Fitzpatrick, J.W., Rosenberg, K.V.: Research priorities for wind energy and migratory wildlife. J. Wildl. Manag. **76**(3), 451–456 (2012). https://doi.org/10.1002/jwmg.327

Avoidance Behaviour of Migrating Raptors Approaching an Offshore Wind Farm

Erik Mandrup Jacobsen, Flemming Pagh Jensen, and Jan Blew

Abstract During three seasons, we studied avoidance behaviour of migrating raptors when approaching an offshore wind farm in northern Baltic Sea 20 km from the coast.

From a substation 1.8 km west of the wind farm, we recorded macro, meso and micro avoidance behaviour of individual raptors approaching the wind farm, using a pre-defined protocol. We defined macro avoidance as when the raptor completely avoids entering the wind farm, meso avoidance as a significant change in altitude or direction before arrival and micro avoidance as a sudden change of flight when passing a turbine at close range.

In total, 466 migrating raptors representing 13 species were observed of which 73% representing 9 species showed macro, meso and/or micro avoidance behaviour.

Macro avoidance was recorded among ten of the species including 59% of red kites, 46% of common kestrels, 42% of sparrowhawks and 30% of honey buzzards. Three quarters of these raptors subsequently left the AOWF in a westerly direction, indicating that they returned to the mainland. The remaining birds flew either north or south parallel to the first row of turbines, suggesting they tried to navigate around the wind farm.

Our study demonstrated a barrier effect from the offshore wind farm influencing the migration of raptors by forcing some birds to use alternative and potentially more risky sea crossings. This may potentially affect survival and fitness of individuals and populations. Accordingly, we recommend that the location of important raptor migration routes is taken into consideration in siting of future offshore wind farms.

Keywords Bird migration · Macro · meso and micro avoidance · Behavioural response · Barrier effect

E. M. Jacobsen (✉) · F. P. Jensen
Orbicon A/S, Taastrup, Denmark
e-mail: emja@orbicon.dk; fpje@orbicon.dk

J. Blew
BioConsult SH GmbH & Co. KG, Husum, Germany
e-mail: j.blew@bioconsult-sh.de

R. Bispo et al. (eds.), *Wind Energy and Wildlife Impacts*,
https://doi.org/10.1007/978-3-030-05520-2_3

1 Introduction

The last decade has witnessed a massive expansion in large offshore wind farms, in particular in the UK, Denmark, Germany, The Netherlands and Belgium [1]. Potentially, offshore wind farms pose a variety of impacts to birds, most notably: (1) collisions with turbines; (2) displacement of birds due to effective loss of habitat; and (3) barrier effects where the wind farm creates an obstacle to regular movements to and from breeding colonies or migration [2–6].

So far, collision has mostly been the focus of raptor studies at wind farms including the long lasting studies of the impact on raptor populations at Altamont, California [7, 8] and in Spain [9].

Here we focus on the barrier effect of an offshore wind farm constructed within a migration corridor of raptors.

Many raptors tend to avoid crossing open sea when migrating [10, 11]. Instead, they follow land areas until a crossing is unavoidable and therefore concentrate at peninsulas or other narrow stretches of land in order to reduce the risks and energy expenditures associated with active flight over water [12].

Avoiding a wind farm during migration prevents the risk of collisions with the turbines but at the cost of a longer flight route. Obviously, the extra cost of flying around one wind farm is negligible, but the cumulative effect of avoidance behaviour at many wind farms along the flyway could potentially increase the risk for more important and long-term consequences [13]. Furthermore, if the barrier created by an offshore wind farm forces the migrating raptors to turn back and instead use a longer sea crossing in another place, it could lead to significant additional energy expenditure.

At the peninsula Djursland in western Denmark, migrating raptors concentrate each spring [14] before crossing the Kattegat (northernmost part of the Baltic Sea) on their way to Sweden. When leaving the coast at Djursland, the raptors typically take an easterly or northeasterly direction and first cross 45 km of open sea to reach the small island Anholt before continuing for another 50 km to the west coast of Sweden (Fig. 1).

In 2013, a large offshore wind farm was constructed halfway between Djursland and Anholt (20 km off the Djursland coast). Anholt Offshore Wind Farm (AOWF) consists of 111 large 3.6 MW turbines arranged in up to 20 km long rows perpendicularly to the flightpath used by the migrating raptors leaving Djursland. The distance between the individual turbines is between 500 and 800 m (Fig. 1).

During three spring seasons, we studied the potential barrier effect of the AOWF on migrating raptors by quantifying macro, meso and micro avoidance behaviour of individual birds as they approached the turbines.

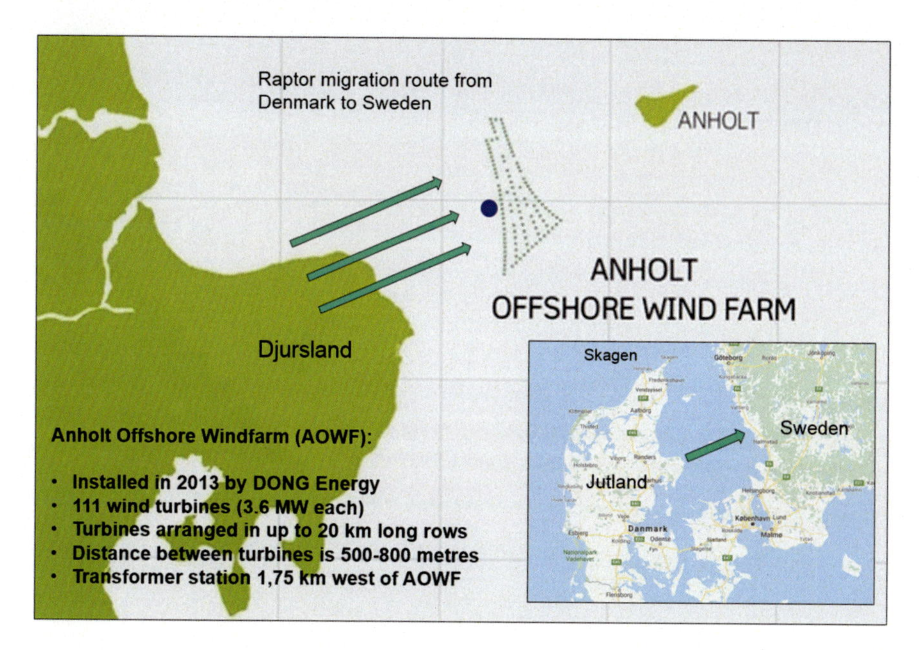

Fig. 1 Direction of raptors migration from Djursland across the Kattegat (Baltic Sea) to Sweden in spring and the AOWF situated halfway between the coast and the island Anholt. The blue dot indicates the position of the substation used for data collection in this study

Table 1 Observation periods in spring 2014–2016

2014	2015	2016
20–24 March	7–10 March	30 March–4 April
28–31 March	26–28 March	12–16 April
10–13 April	8–11 April	29 April–2 May
23–27 April	22–25 April	7–11 May
7–10 May	8–12 May	18–22 May
21–23 May	20–24 May	27–31 May
1–4 June	1–5 June	

2 Methods

2.1 Observations

In spring 2014, 2015 and 2016, we carried out 89 days of standardized systematic recordings of behavioural responses of migrating raptors approaching the AOWF (Table 1). The observations were made from an offshore substation (Fig. 2) located 1.8 km west of the nearest turbine.

Between 7 and 12 h of observations were made each day from early morning (6–7 am) until mid-afternoon (when no more migrating raptors were observed). The

Fig. 2 The substation used for recording behavioural reaction of migrating raptors. The closest turbines are seen in the background (c. 1.8 km away)

observations were only carried out on days with good visibility (min. 2–3 km), little or no precipitation and low wind speed (<6 m/s).

Observations were made using handheld binoculars (10× magnification) and 30× binoculars mounted on a tripod. To record the movements of raptors flying parallel to the row of turbines outside the range of the binoculars, a horizontally mounted Furuno surveillance radar was used with automatic storage of screenshots of the radar screen every minute.

With a visibility of at least 2–3 km, the migrating raptors could be followed all the way to the first row of turbines and sometimes onwards in between the turbines. When approaching the first row of turbines, the behaviour of the migrating raptor was recorded according to the following pre-defined protocol:

1. Is the raptor avoiding the wind farm by hesitating and starting to circle or flying parallel to the row of turbines or even turning back?
2. What is the altitude and direction of the raptor when it arrives to the wind farm? Examples: it is "flying NE twice the height of a turbine, under the swept area; there is no visible change in altitude or direction etc."?
3. Is the bird changing the flight altitude or direction when passing the first turbines (is the bird for example gaining height and fly over the swept area/losing altitude to fly under)?
4. Is the raptor taking a path between the turbines?

5. Is the raptor (apparently) ignoring the turbines and flying very close or through the swept area?
6. Are any close-range ("last moment") evasive movements visible?

The collected behavioural data range from a few records of the flight altitude and direction close to a turbine to observations lasting more than 40 min when, for example, the birds first hesitated and afterwards were flying parallel to the turbine row before eventually passing the wind farm (or was turning back).

2.2 Data Processing

For the purpose of this study, we define macro, meso and micro avoidance as follows:

- *Macro avoidance*: the raptor avoids entering the wind farm by either turning back or flying parallel to the first turbine row.
- *Meso avoidance*: a significant change in altitude or direction before arriving to the first row of turbines (after which the raptor enters the wind farm).
- *Micro avoidance*: a sudden change in the flight pattern when passing the first turbine at close range.

From the number of observed raptors approaching the AOWF, we calculated for each species the number and percentage that showed macro, meso and micro avoidance behaviour.

3 Results

In total 466 migrating raptors representing 13 different species were observed approaching the AOWF from mainly western or southwesterly directions (Table 2). Of these 340 individuals (73%) representing 9 species showed macro, meso and/or micro avoidance behaviour (Table 2).

The percentage of the different avoidance types for each species is shown in Table 3. It should be noted that the numbers listed under macro avoidance include birds that did not enter the wind farm at all but either left the AOWF in a westerly direction, indicating that they were returning to the mainland (75%), or flew north or south parallel to the first row of turbines (25%), suggesting they were trying to navigate around the wind farm.

Table 2 Total number of observed raptors approaching the AOWF and the numbers and percentages showing macro, meso and/or micro avoidance according to our definitions (see text)

Species	Number observed ($n = 466$)	Number showing macro, meso and/or micro avoidance	% showing avoidance behaviour
Honey buzzard (*Pernis apivorus*)	27	17	63
Red kite (*Milvus milvus*)	32	26	81
White-tailed eagle (*Haliaeetus albicilla*)	1	0	0
Marsh harrier (*Circus aeruginosus*)	29	19	66
Hen harrier (*Circus cyaneus*)	7	4	57
Sparrowhawk (*Accipiter nisus*)	119	101	85
Goshawk (*Accipiter gentilis*)	1	0	0
Common buzzard (*Buteo buteo*)	195	133	68
Rough-legged buzzard (*Buteo lagopus*)	2	0	0
Osprey (*Pandion haliaetus*)	15	12	80
Common kestrel (*Falco tinnunculus*)	22	21	96
Merlin (*Falco columbarius*)	14	7	50
Peregrine falcon (*Falco peregrinus*)	2	0	0

Table 3 Percentage of raptors showing macro, meso and/or micro avoidance behaviour when arriving to the AOWF

Species	% showing macro avoidance	% showing meso avoidance	% showing micro avoidance
Honey buzzard (*Pernis apivorus*)	30	33	0
Red kite (*Milvus milvus*)	59	22	0
Marsh harrier (*Circus aeruginosus*)	21	38	7
Hen harrier (*Circus cyaneus*)	29	14	14
Sparrowhawk (*Accipiter nisus*)	42	37	6
Common buzzard (*Buteo buteo*)	27	41	1
Osprey (*Pandion haliaetus*)	13	60	7
Common kestrel (*Falco tinnunculus*)	46	50	0
Merlin (*Falco columbarius*)	14	36	0

4 Discussion

Studies of avoidance behaviour of birds around offshore wind farms have so far mainly focused on seabirds [15] although a general avoidance of raptor abundance when (onshore) wind farms are constructed is well known, for example [16]. Strong avoidance of migrating raptors to onshore wind farms has also been documented [17, 18], but a recent Danish study found soaring raptors on migration when leaving the coast displayed a significant attraction behaviour towards an offshore wind farm 12 km away [19]. It is suggested that the most likely driver behind this attraction is an "island effect" [19].

Our findings suggest avoidance behaviour among nearly all raptor species when getting close to the offshore wind farm. With the exception of four species which were only recorded in very low numbers, between half and nearly all the individuals of the different migrating raptor species showed at least one of the three avoidance behaviour types. Most striking was probably the large percentage of some species that completely avoided entering the wind farm and turned back.

The attraction of a distant offshore wind by raptors leaving the coast as shown by [19] suggests that they respond depending on how far they are from the offshore wind farms. Initially the birds are attracted perhaps because they believe the wind farm is an island which could act as a stepping stone during their sea crossing, but as they get closer, they realize that this is not the case and many then change course or turn back. This is supported by the attraction being inversely related to the distance to the wind farm [19].

Because distant offshore wind farms may attract migrating raptors, it has been suggested that this potentially makes the birds far more at risk of colliding with wind turbines at sea than previously assumed [19]. This appear not always to be the case since our findings show that once the migrating raptors get close to the turbines, large numbers avoid the wind farm and are therefore not at risk of collision.

While the increased energy expenditure associated with flying round the wind farm and then continuing towards Anholt and Sweden is probably minor to the raptors in question, there could be significant risks associated with sea crossings in other locations. This is most likely not the case for the raptors that turned back from the AOWF. These raptors will probably continue north over land and make the sea crossing from one of the other well-known migration spots in Denmark, in particular from the northernmost tip of Jutland (at Skagen) where the distance to Sweden is about 55 km. However, in other parts of these birds' range, the turning back from a favoured sea crossing route could lead to a much longer and more risky sea crossing with significant increased energy expenditure.

5 Conclusions

Our study showed a barrier effect of an offshore wind farm influencing the migration of raptors by forcing many of the birds to use other and potentially more risky alternative sea crossings. This impact may potentially affect the survival and fitness of individuals and populations.

When siting future offshore wind farms, we therefore recommend planners to take into consideration the location of important migration routes for raptors.

Acknowledgements The post-construction monitoring study at the AOWF was financed by Ørsted, Denmark (Previously DONG Energy). The photo was kindly provided by Lars Maltha Rasmussen.

References

1. Colmenar-Santos, A., Perera-Perez, J., Borge-Diez, D., de Palacio-Rodríguez, C.: Offshore wind energy: a review of the current status, challenges and future development in Spain. Renew. Sustain. Energy Rev. **64**, 1–18. Elsevier (2016)
2. Drewitt, A.L., Langston, R.H.W.: Assessing the impacts of wind farms on birds. Ibis. **148**, 29–42 (2006)
3. Humphreys, E.M., Cook, A.S.C.P., Burton, N.H.K.: Collision, Displacement and Barrier Effect Concept Note BTO Research Report No. 669. The British Trust for Ornithology, The Nunnery, Thetford (2015)
4. Kuvlesky, W.P., Brennan, L.A., Morrison, M.L., Boydston, K.K., Ballard, B.M., Bryant, F.C.: Wind energy development and wildlife conservation: challenges and opportunities. J. Wildl. Manag. **71**, 2487–2498 (2007)
5. Stewart, G.B., Pullin, A.S., Coles, C.F.: Poor evidence-base for assessment of windfarm impacts on birds. Environ. Conserv. **34**, 1–11 (2007)
6. Rydell, J., Engström, H., Hedenström, A., Kyed Larsen, J., Pettersson, J., Green, M.: The Effect of Wind Power on Birds and Bats – A Synthesis, 152 pp. Swedish Environmental Protection Agency, Stockholm (2012)
7. Orloff, S., Flannery, A.: Wind turbine effects on avian activity, habitat use and mortality in Altamont Pass and Solano County wind resource areas, 1989–1991. Unpublished final report prepared by BioSystems Analysis Inc., Tiburon, California, for the California Energy Commission, Sacramento, grant 990-89-003. 199 p (1992)
8. Smallwood, K.S., Thelander, C.: Bird mortality in the Altamont Pass Wind Resource Area, California. J. Wildl. Manag. **72**, 215–223 (2008)
9. Carrete, M., Sanches-Zapata, J.A., Benitez, J.R., Lobon, M., Donazar, J.A.: Large scale risk-assessment of wind-farms on population viability of a globally endangered raptor. Biol. Conserv. **142**, 2954–2961 (2009)
10. Alerstam, T.: Analysis and theory of visible bird migration. Oikos. **30**, 273–349 (1978). https://doi.org/10.2307/3543483
11. Newton, I.: Bird Migration. Collins, London (2010)
12. Alerstam, T.: Bird Migration. Cambridge University Press, Cambridge (1990)
13. Masden, E.A., Hayden, D.T., Fox, A.D., Furness, R.W., Bullman, R., Desholm, M.: Barriers to movement: impacts of wind farms on migrating birds. J. Mar. Sci. **66**, 746–753 (2009)
14. BirdLife International, Denmark. Homepage, https://dofbasen.dk. Last accessed 28 Mar 2018
15. Krijgsveld, K.L.: Avoidance behaviour of birds around offshore wind farms. Overview of knowledge including effects of configuration. Bureau Waardenburg bv. 35 pp (2014)
16. Garvin, J.C., Jennelle, C.S., Drake, D., Grodsky, S.M.: Response of raptors to a windfarm. J. Appl. Ecol. **48**, 199–209 (2011)
17. Villegas-Patraca, R., Cabrera-Cruz, S.A., Herrera-Alsina: Soaring migratory birds avoid wind farm in the Isthmus of Tehuantepec, Southern Mexico. PLoS ONE. **9**(3), e92462 (2014). https://doi.org/10.1371/journal.pone.0092462
18. Cabrera-Cruz, S.A., Villegas-Patraca, R.: Response of migrating raptors to an increasing number of wind farms. J. Appl. Ecol. **53**, 1667–1675 (2016)
19. Skov, H., Desholm, M., Heinänen, S., Kahlert, J.A., Laubek, B., Jensen, N.E., Zydelis, R., Jensen, B.P.: Patterns of migrating soaring migrants indicate attraction to marine wind farms. Biol. Lett. **12**, 20160804 (2017). https://doi.org/10.1098/rsbl.2016.0804

Estimating Potential Costs of Cumulative Barrier Effects on Migrating Raptors: A Case Study Using Global Positioning System Tracking in Japan

Dale M. Kikuchi, Toru Nakahara, Wataru Kitamura, and Noriyuki M. Yamaguchi

Abstract Wind farms along the migration route of birds act as unnatural barriers, and avoiding them during flight may require the expenditure of extra energy. Information regarding cumulative effects of barriers on migrating birds is generally lacking, mainly because of the complexities of monitoring the number of encounters of migratory birds with wind farms and their flight path for avoiding these barriers. It would be desirable to develop a general method for monitoring the rate at which migratory birds encounter wind farms. In this study, we attempted to assess the potential cumulative barrier effects on 17 eastern buzzards (*Buteo japonicus*) and eight Oriental honey-buzzards (*Pernis ptilorhynchus*) using global positioning system (GPS) tracking data. We obtained the location data of wind turbines in Japan and migration paths of the birds using GPS loggers and assumed four scenarios that birds could use to avoid the wind turbines along their routes. Although the number of studied individuals was limited and the impact of the cumulative effects are inconclusive at the present stage, the estimated additional distance, time, and energetic cost during one migration were no more than 31.97 km, 75.74 min and 132 kJ, respectively, which were relatively small. Additionally, we showed the possibility that GPS tracking could provide information on migration episodes of birds

---

Authors Dale M. Kikuchi, Toru Nakahara have been equally contributed to this chapter.

D. M. Kikuchi (✉) · W. Kitamura
Tokyo City University, Kanagawa, Japan
e-mail: dale@tcu.ac.jp; kitamura@tcu.ac.jp

T. Nakahara
Graduate School of Fisheries and Environmental Sciences, Nagasaki University, Nagasaki, Japan

Kitakyushu Museum of Natural History and Human History, Fukuoka, Japan
e-mail: nakahara_t@kmnh.jp

N. M. Yamaguchi
Graduate School of Fisheries and Environmental Sciences, Nagasaki University, Nagasaki, Japan
e-mail: noriyuki@nagasaki-u.ac.jp

R. Bispo et al. (eds.), *Wind Energy and Wildlife Impacts*,
https://doi.org/10.1007/978-3-030-05520-2_4

associated with wind farms and could be a promising method for addressing the cumulative barrier effects. We expect that a higher sampling frequency of locations would enable precise measurement of avoidance flights and energetic costs.

Keywords Cumulative barrier effects · Global positioning system · Migrating raptors

1 Introduction

Wind farms act as unnatural barriers to the movement of birds. For example, radar survey revealed that migrating birds avoid entering wind farms by changing their flight paths [1–3]. Altering flight paths may represent an extra energetic cost and potentially affect the body condition, which is associated with survival and reproductive success [1, 4]. Although the energetic cost of avoiding an individual wind farm may have small effects, encounters with additional wind farms along the migration route may have cumulative and significant effects [1]. However, appropriate methods for monitoring whole migration paths and cumulative barrier effects are lacking. This is because radar survey has limited abilities to track a part of the migration path and assess the barrier effects at a particular site. Thus, the information on the number of encounters with wind farms during migration is unclear. Additionally, surveys using radars are mostly limited to offshore wind farms (reviewed in Cabrera-Cruz and Villegas-Patraca [3]) because the radio waves can diffract around obstacles, such as mountains. Therefore, it would be desirable to develop a general method of monitoring the rate at which migratory birds encounter wind farms both onshore and offshore. This may facilitate the assessment of additional flight distance, time, and energetic cost during migration, which might be associated with avoidance of wind farms.

Advances in animal tracking technology, including miniaturized onboard global positioning system (GPS) devices, provide information on the long-term movement path of animals [5]. As compared with surveys using radars, GPS tracking can track a lesser number of individuals. However, GPS tracking has a global range of operation, and it is possible to solve the problem of trackable range, which is a constraint of radar-based surveys [6]. Therefore, GPS tracking may resolve the current limitations associated with counting the number of encounters of migrating birds with wind farms. This behavioral information would facilitate the assessment of the cumulative barrier effects of wind farms both onshore and offshore.

In the present study, we tested the effectiveness of GPS tracking to assess the potential cumulative barrier effects on two raptor species, the eastern buzzard (*Buteo japonicus*) and the Oriental honey-buzzard (*Pernis ptilorhynchus*). We obtained migration paths of the birds and assumed four scenarios by which the birds circumvented wind turbines along their routes. Thereafter, we estimated the additional flight distance, time, and energetic costs associated with the avoidance of wind farms.

2 Materials and Methods

2.1 Capture and GPS Tracking

Eastern buzzards mainly breed in the Chubu region and north of this region in Japan and spend the winter in most areas of Japan [7]. Oriental honey-buzzards also breed in Japan and overwinter in Southeast Asia [8]. Therefore, many individuals of these species pass along the Japan archipelago during the spring and autumn migrations. We tracked 17 eastern buzzards and 8 Oriental honey-buzzards. Eastern buzzards were captured from December 2016 to February 2017 at wintering sites in Japan: Fukuoka Prefecture (seven individuals), Hyogo Prefecture (five individuals), and Nagasaki Prefecture (five individuals). All Oriental honey-buzzards were captured in June 2017 at breeding sites in the Aomori Prefecture, Japan.

We captured these birds harmlessly using clap net traps licensed by Fukuoka, Hyogo, Nagasaki, and Aomori Prefectures in Japan and the Ministry of the Environment, Government of Japan. We measured their body weights (kg) and photographed each bird with its wings expanded. From the images, we calculated the wingspan (m) and wing area (m^2) using ImageJ ver.1.51k (https://imagej.nih.gov/ij/, see Appendix 1).

The birds were tracked using GPS loggers, WT-300s standard type (32 g, KoEco) for the 12 eastern buzzards and 8 Oriental honey-buzzards, and Harrier-M (15 g, Ecotone) for 5 eastern buzzards (Fig. 1, see also Appendix 1). We obtained the date and time as well as the longitude and latitude of the locations using the loggers. The loggers were programmed to record GPS locations every 30 min if light condition was good; otherwise, the locations were recorded every 4 h on the WT-300s and every 20 min on the Harrier-M. However, they sometimes failed to record the information at constant time intervals.

Fig. 1 Image of a tagged bird (Oriental honey-buzzard). (Dale M. Kikuchi owns the copyrights)

2.2 Calculation of Additional Distance and Time in a Scenario Avoiding Wind Turbines

We counted the number of wind turbines and wind farms found along the migration line in Japan. Information on the locations of wind turbines was obtained from two websites of the Ministry of Land, Infrastructure, Transport and Tourism (http://nlftp.mlit.go.jp/ksj/index.html and http://www.cab.mlit.go.jp/tcab/aerial_beacon/01.html). The migratory route was regarded as a line created by connecting consecutive GPS locations. Estimation of avoidance distance was based on the avoiding behavior recorded in previous studies. They showed that raptor species, including *Buteo* sp., and the Oriental honey-buzzard avoid wind turbines or wind farms from approximately 2000 m [3, 9, 10]. Therefore, we prepared multiple scenarios of avoidance. We created 100, 500, 1000, and 2000 m radius buffers around the wind turbines using QGIS ver. 2.18.9 (https://qgis.org/ja/site/) and counted the number of buffers that each bird passed. When the wind turbines were in a row within 500 m intervals, we regarded them as one wind farm and counted them accordingly.

In addition, we examined the potential and additional distances and times assuming a scenario that the birds circumvented buffers around wind turbines, although there was no evidence that birds return to an original route after showing avoidance behavior. To estimate the additional distance covered under different scenarios, the distance of a migration route within a buffer was measured at each wind farm/turbine. The potential cumulative additional distance (*CAD*) was calculated according to Eq. (1):

$$\mathrm{CAD} = \sum_{i=1}^{n} (D_{\mathrm{ARi}} - D_{\mathrm{ORi}}) \tag{1}$$

where D_{ORi} is the distance of original route (km), which was the distance of original migration route within the *i*th buffer, and D_{ARi} is the distance of avoidance route (km), which was the distance of an arc between intersections of the original migration route of a bird and circumference of the *i*th buffer (Fig. 2). When the GPS location was within the buffer area, we calculated the D_{ARi} by assuming the bird flew to the circumference of the *i*th buffer at the shortest distance and then returned to an original migration route. We did not take the landscape features into consideration.

The potential cumulative additional time (CAT) was calculated according to Eq. (2):

$$\mathrm{CAT} = \sum_{i=1}^{n} (D_{\mathrm{ARi}} - D_{\mathrm{ORi}}) \times V_{\mathrm{gi}} \tag{2}$$

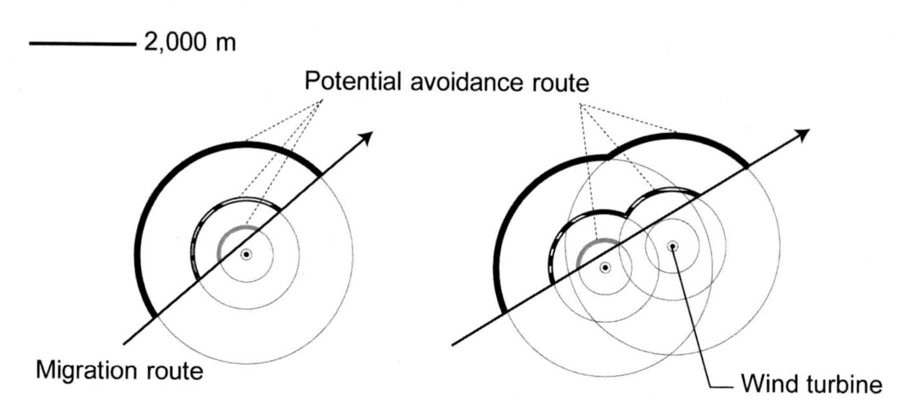

Fig. 2 Conceptual diagram of a scenario where birds circumvent wind turbines. Black, shaded, and gray trajectories represent migration routes assumed when 2000, 1000, and 500 m radius buffers are avoided

where V_{gi} was the ground speed (km/minute) measured from the distance and the time between two GPS locations located near the *i*th buffer. When a bird rested for one night between those two GPS locations, we calculated the time taken to fly the additional distance by assuming the ground speed calculated when the bird started to fly from the point just before the buffer at 7:00 on the next morning. Because we did not know the most appropriate scenario of avoidance, we have presented the range of costs and discussed the worst case below.

2.3 Calculation of Energetic Costs During Migration

We used a biomechanical modeling software for avian flight (Flight 1.25) [11, 12] to estimate the energetic costs to circumvent wind turbines. The biomechanical model is based on a classic theory of fixed-wing aircraft, and it requires morphological information of the birds (body mass, wing area, and wingspan). Using Flight 1.25, we first calculated the chemical power (rate of energy consumption) at the estimated minimum power speed (lowest power required speed) for a steady flapping flight for each individual. Then, we calculated the energy required to fly for a CAT in Japan by multiplying the chemical power at minimum power speed and the additional flight duration. We estimated the energy costs for the whole migration flight by assuming that the birds flew along their migration route at their minimum power speed and by multiplying the total flight duration and chemical power at minimum power speed. It is noteworthy that we could not consider the ratio of gliding and soaring flight because the GPS loggers we used did not provide information on the flight mode of the birds. Gliding and soaring flight are common flight modes of raptors that extract energy from gravity, vertical airflow such as thermal and ridge lift, or both to support their weight in the air [11]. Therefore, our assumption may have overestimated the energy consumption of the flight.

3 Results

3.1 Spring Migration of Eastern Buzzards

The distance traveled during the spring migration of the eastern buzzards was from 373 to 3511 km ($n = 17$, see Appendix 2). Five individuals did not pass the vicinity of the wind turbines or wind farms, but the remaining 12 individuals passed within 1000 m of the wind turbines and farms (Fig. 3a, see also Appendix 2). Six individuals passed within 100 m of the wind turbines and farms (Fig. 3a, see also Appendix 2).

When we assumed that birds circumvented the buffers around wind turbines, the additional distance they needed to travel increased by up to 31.97 km (Fig. 3b). It was comparable to 1.14% of the entire distance of the spring migration. The additional time required to travel ranged from 0 to 75.7 min (Fig. 3c). The estimated additional energetic cost ranged from 0 to 132 kJ (Fig. 3d), which corresponds to a 0–2.5% increase in the entire energetic cost of the spring migration (Fig. 3e, see also Appendix 2).

3.2 Autumn Migration of Eastern Buzzards

The distance traveled during the autumn migration of the eastern buzzards was from 1451 to 2444 km ($n = 7$, see also Appendix 3). Only one individual did not pass the vicinity of the wind turbines or wind farms, but the remaining six passed within 1000 m of the wind turbines and farms (Fig. 3a, see also Appendix 3). Three individuals passed within 100 m of the wind turbines and farms (Fig. 3a, see also Appendix 3).

When we assumed that the birds circumvented buffers around the wind turbines, the additional distance they were required to travel increased by up to 15.14 km (Fig. 3b). It was comparable to 0.62% of the entire distance of the spring migration. The additional time required to travel ranged from 0 to 43.5 min (Fig. 3c). The estimated additional energetic cost ranged from 0 to 93 kJ (Fig. 3d), which corresponds to a 0–1.2% increase in the entire energetic cost of the spring migration (Fig. 3e, see also Appendix 3).

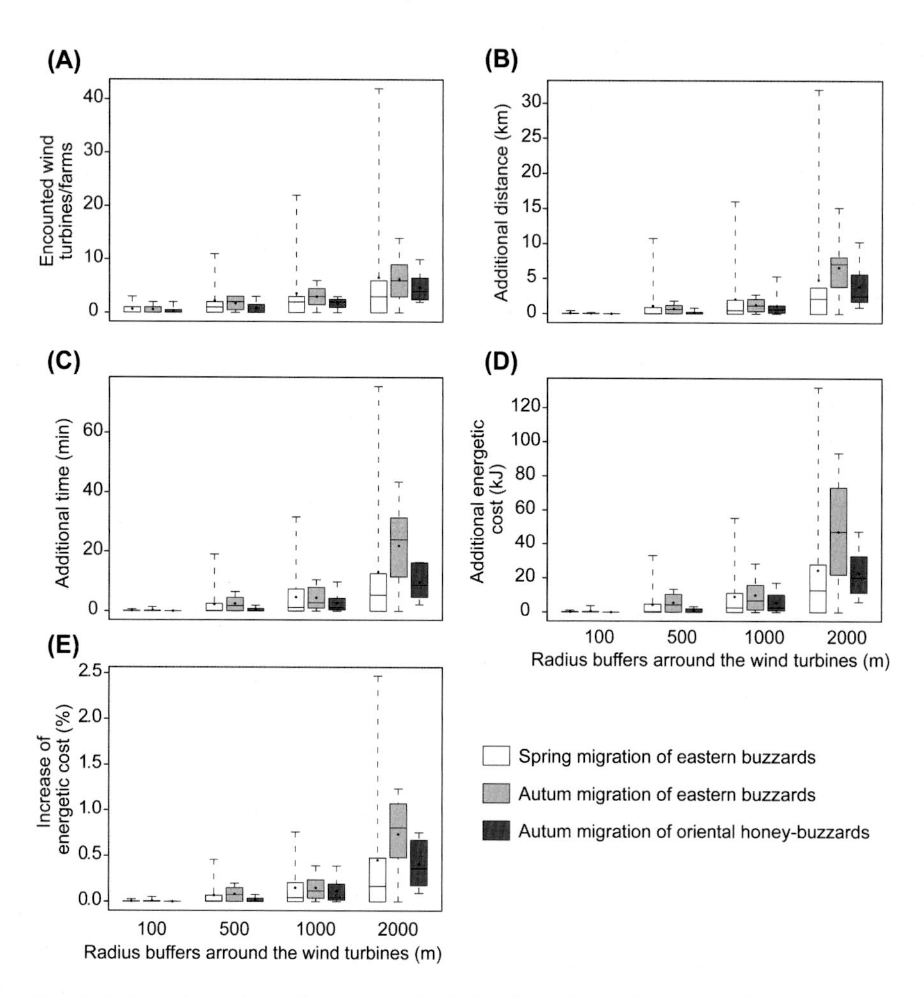

Fig. 3 Estimated numbers of encountered wind turbines/farms (**a**). The estimated potential increase of flying distance (**b**) and time (**c**) assuming that a bird circumvents buffers. Estimated additional energetic costs associated with avoidance of wind turbines and wind farms (**d**) and percentage of the total migration flight (**e**). Boxes represent the lower quartile, median, and upper quartile values. The whiskers extending above and below the box show the highest and lowest values, respectively. The filled circles show the mean value

3.3 Autumn Migration of Oriental Honey-Buzzards

The distance traveled in Japan during the autumn migration of the Oriental honey-buzzards was from 1524 to 1888 km ($n = 8$; see Appendix 4). All individuals passed within 2000 m of the wind turbines and farms (Fig. 3a; see also Appendix 4). Two individuals passed within 100 m of wind turbines and farms (Fig. 3a; see also Appendix 4).

When we assumed that the birds circumvented the buffers around wind turbines, the additional distance they needed to travel increased by up to 10.28 km (Fig. 3b). It was comparable to 0.61% of the entire distance of the autumn domestic migration in Japan. The additional time required to travel ranged from 0 to 47.5 min (Fig. 3c). The estimated additional energetic cost ranged from 0 to 93 kJ (Fig. 3d), which corresponds to a 0–0.8% increase in the entire energetic cost of the spring migration (Fig. 3e, see also Appendix 4).

4 Discussion

Here, we presented the results of our test case of a strategy for the assessment of cumulative barrier effects of onshore wind farms on migrating raptors using GPS tracking data. The potential maximum increase in distance and flight time for avoiding wind farms was 31.97 km and 75.74 min for eastern buzzards and 10.28 km and 16.44 min for the Oriental honey-buzzards, respectively (Fig. 3). The potential additional energetic costs were up to 132 and 93 kJ, which is equivalent to 3.5 and 2.5 g (0.45% and 0.28% of the body mass) of fat (37.7 kJ/g) for the eastern and Oriental honey-buzzards, respectively. Although we cannot conclude the most reliable scenario, these results appeared to suggest that the wind turbines and farms along the migratory routes in Japan did not largely alter the flight costs of the individual birds studied. However, the cumulative barrier effects on the studied species are still inconclusive because the number of studied individuals was limited. We could not account for undetermined migration routes. If there were cases where raptors encountered more wind farms, this study might have underestimated the cumulative barrier effects. Moreover, it would be preferable to continue the present survey to monitor the effect of pre- and post-construction stages of new wind farms to evaluate the effects on their migration routes [3].

Our methods are based on several assumptions, and there are two major limitations at the present stage of this study. First, the flight path was defined as the horizontal line connecting the GPS positions although the behavior of the birds during that interval is unknown. This is because the GPS locations were obtained at a minimum interval of 20 min and its altitude data were not considered for analysis. The flight paths are considerably complicated in three dimensions, and we are not certain whether the birds were approaching or avoiding the wind farms during the intervals. Therefore, a higher sampling frequency of locations and precise measurement of altitude would be desirable. However, there is a trade-off between battery consumption and sampling frequency of GPS devices [13]. Therefore, GPS devices with low power consumption should be designed to increase the recording period further.

Second, the calculation of the energetic cost was based on the classic theory of a fixed-wing aircraft, and we did not account for wind effects and flight mode (i.e., flapping, gliding, and soaring) as mentioned in the Sect. 2. Wind speed strongly affects the ground speed of flying birds, assuming that it has slight effects on airspeed and power requirement of flight [11]. For example, headwind decreases the ground speed of flight and increases the energetic costs at a given distance [11]. Therefore, it is necessary to account for the wind directions for a more precise prediction. Alternatively, to estimate the energetic cost of the flight, it would be ideal to measure the heart rate of the birds, as a proxy of the energy consumption rate, using a heart rate sensor [14]. Therefore, simultaneous recording of the GPS locations and heart rate would be one of the best strategies for assessing the cumulative barrier effects.

It appears that GPS tracking is a promising method for assessing the cumulative barrier effects of wind farms on migratory birds. Although limitations remain to be addressed, we showed that GPS tracking could provide information on the migration episodes of birds associated with wind farms. For the practical use of GPS tracking for the assessment of cumulative barrier effects, it would be necessary to include a large number of individuals and species in the study group. This is because birds often use a variety of migration routes and exhibit different types of avoidance behaviors owing to wind farms. Moreover, long-term monitoring of GPS tracking would make it possible to compare the bird migrations before and after the construction of wind farms. In the future, widening the scope of this approach to a variety of migratory species could contribute to improving the spatial planning of wind farms to resolve the conflicts between migratory birds and wind farms.

Acknowledgments We would like to thank Fumihito Nakayama, Naoya Hijikata, Fumitaka Iseki, Fumio Katsuno, Kazuhisa Oue, Takashi Suzumegano, and Asuka Hirano for the field assistant and support. This paper is based on results obtained from a project commissioned by the New Energy and Industrial Technology Development Organization (NEDO).

Appendices

Appendix 1 (Tables A1 and A2)

Table A.1 Measurement of captured birds and power required for flapping flight of eastern buzzard

Bird ID (tracker)	Capture site	Wing span (m)	Wing area (m^2)	Weight (kg)	Minimum chemical power (W)
ns1601 (WT-300s)	Fukuoka	1.17	0.21	1.01	45.1
ns1602 (WT-300s)	Fukuoka	1.25	0.24	0.90	34.5
ns1603 (WT-300s)	Fukuoka	1.29	0.26	0.99	38.5
ns1604 (WT-300s)	Fukuoka	1.20	0.25	1.00	45.5
ns1606 (WT-300s)	Fukuoka	1.16	0.21	0.90	37.9
ns1607 (WT-300s)	Fukuoka	1.19	0.22	0.96	40.6
ns1608 (WT-300s)	Hyogo	1.22	0.24	0.94	38.2
ns1610 (WT-300s)	Fukuoka	1.21	0.23	0.91	36.7
ns1612 (WT-300s)	Hyogo	1.24	0.23	1.11	47.4
ns1614 (WT-300s)	Hyogo	1.22	0.23	1.01	42.2
ns1615 (WT-300s)	Hyogo	1.23	0.23	0.96	38.7
ns1619 (WT-300s)	Hyogo	1.19	0.22	0.90	37.0
HAR20 (Harrier-M)	Nagasaki	1.25	0.23	0.88	31.9
HAR21 (Harrier-M)	Nagasaki	1.18	0.21	0.77	29.1
HAR22 (Harrier-M)	Nagasaki	1.24	0.25	0.69	23.5
HAR23 (Harrier-M)	Nagasaki	1.28	0.25	0.79	27.2
HAR24 (Harrier-M)	Nagasaki	1.23	0.25	0.90	35.8

Table A.2 Measurement of captured birds and power required for flapping flight of Oriental honey-buzzard

Bird ID (tracker)	Capture site	Wing span (m)	Wing area (m^2)	Weight (kg)	Minimum chemical power (W)
ngsku1601 (WT-300s)	Aomori	1.37	0.32	1.19	48.9
ngsku1605 (WT-300s)	Aomori	1.38	0.30	0.94	32.6
ngsku1701 (WT-300s)	Aomori	1.45	0.33	1.16	42.7
ngsku1704 (WT-300s)	Aomori	1.46	0.33	1.22	45.2
ngsku1706 (WT-300s)	Aomori	1.31	0.27	0.84	29.1
ngsku1707 (WT-300s)	Aomori	1.33	0.32	1.13	48.7
ngsku1709 (WT-300s)	Aomori	1.33	0.26	1.02	38.0
ns1621 (WT-300s)	Aomori	1.41	0.29	1.01	34.4

Appendix 2 (Tables A3, A4, and A5)

Table A.3 Distance traveled, number of GPS positioning points, and estimated total cost during spring migration of eastern buzzards

Bird ID	Distance traveled (km)	Number of GPS positioning points	Estimated total cost ($\times 10^6$ kJ)
ns1601	2085	709	7.58
ns1602	2516	715	7.55
ns1603	1315	256	4.33
ns1604	1891	348	7.05
ns1606	1256	196	4.00
ns1607	3511	1354	11.9
ns1608	1323	968	4.28
ns1610	1243	419	3.90
ns1612	373	29	1.43
ns1614	802	347	2.80
ns1615	675	183	2.20
ns1619	963	158	3.02
HAR20	2503	2234	7.00
HAR21	2800	2639	7.28
HAR22	1472	717	3.26
HAR23	2423	1604	6.05
HAR24	2454	1423	7.57

Table A.4 Potential number of wind turbines and farms and increased distance to circumvent them during spring migration of eastern buzzards

	Number of buffers around wind turbines and wind farms on a migration route				Increase of distance traveled assuming a bird circumvents buffers (km)			
	Buffer size (radius)				Buffer size (radius)			
Bird ID	100 m	500 m	1000 m	2000 m	100 m	500 m	1000 m	2000 m
ns1601	0	0	1	6	0	0	0.05	2.16
ns1602	0	0	1	1	0	0	0.20	1.14
ns1603	0	1	2	3	0	0.03	0.62	2.56
ns1604	0	1	2	2	0	0.06	0.49	2.49
ns1606	0	0	0	0	0	0	0	0
ns1607	3	7	10	15	0.16	2.93	4.30	10.16
ns1608	0	0	1	3	0	0	0.03	1.24
ns1610	0	0	0	0	0	0	0	0
ns1612	0	0	0	0	0	0	0	0
ns1614	0	0	0	0	0	0	0	0
ns1615	0	0	0	0	0	0	0	0
ns1619	1	2	3	3	0.03	0.91	1.98	3.76
HAR20	1	2	2	2	0.28	1.14	1.61	2.50
HAR21	2	11	22	42	0.17	10.78	16.02	31.97
HAR22	3	10	11	24	0.42	1.66	5.15	16.82
HAR23	0	1	3	5	0	0.68	1.19	2.04
HAR24	1	2	3	6	0.06	0.90	3.95	4.88

Table A.5 Potential costs to circumvent wind turbines and farms during spring migration of eastern buzzards

	Increase of time traveled assuming a bird circumvents buffers (minute)				Increase of energetic cost to travel during additional duration (kJ)			
	Buffer size (radius)				Buffer size (radius)			
Bird ID	100 m	500 m	1000 m	2000 m	100 m	500 m	1000 m	2000 m
ns1601	0	0	0.10	4.73	0	0	0.27	12.80
ns1602	0	0	0.49	2.77	0	0	1.01	5.73
ns1603	0	0.20	3.06	11.41	0	0.46	7.07	26.36
ns1604	0	0.14	1.23	6.45	0	0.38	3.36	17.61
ns1606	0	0	0	0	0	0	0	0
ns1607	0.39	7.01	10.14	23.32	0.95	17.08	24.70	56.81
ns1608	0	0	0.19	5.93	0	0	0.44	13.59
ns1610	0	0	0	0	0	0	0	0
ns1612	0	0	0	0	0	0	0	0
ns1614	0	0	0	0	0	0	0	0
ns1615	0	0	0	0	0	0	0	0
ns1619	0.14	3.80	7.41	12.6	0.31	8.44	16.45	27.97
HAR20	0.61	2.46	3.46	5.39	1.17	4.71	6.62	10.32
HAR21	0.12	19.06	31.71	75.74	0.21	33.28	55.37	132.24
HAR22	0.64	2.53	7.96	57.15	0.90	3.57	11.22	80.58
HAR23	0	0.93	1.62	2.78	0	1.52	2.64	4.54
HAR24	0.15	2.32	12.07	14.00	0.32	4.98	25.93	30.07

Appendix 3 (Tables A6, A7, and A8)

Table A.6 Distance traveled, number of GPS positioning points, and estimated total cost during autumn migration of eastern buzzards

Bird ID	Distance traveled (km)	Number of GPS positioning points	Estimated total cost ($\times 10^6$ kJ)
ns1601	1996	247	7.26
ns1607	1733	102	5.86
HAR20	2223	1322	6.22
HAR21	1924	1130	5.00
HAR22	1451	632	3.22
HAR23	1950	1153	4.87
HAR24	2444	869	7.54

Table A.7 Potential number of wind turbines and farms and increased distance to circumvent them during autumn migration of eastern buzzards

	Number of buffers around wind turbines and wind farms on a migration route				Increase of distance traveled assuming a bird circumvents buffers (km)			
	Buffer size (radius)				Buffer size (radius)			
Bird ID	100 m	500 m	1000 m	2000 m	100 m	500 m	1000 m	2000 m
ns1601	1	2	3	5	0.17	0.62	1.52	6.54
ns1607	1	3	5	11	0.09	0.72	1.18	7.80
HAR20	0	3	4	6	0	1.70	2.68	8.44
HAR21	0	0	0	0	0	0	0	0
HAR22	0	0	1	1	0	0	0.25	1.17
HAR23	0	1	2	7	0	0.02	0.48	7.09
HAR24	2	3	6	14	0.04	1.83	2.74	15.14

Table A.8 Potential costs to circumvent wind turbines and farms during autumn migration of eastern buzzards

Bird ID	Increase of time traveled assuming a bird circumvents buffers (minute)				Increase of energetic cost to travel during additional duration (kJ)			
	Buffer size (radius)				Buffer size (radius)			
	100 m	500 m	1000 m	2000 m	100 m	500 m	1000 m	2000 m
ns1601	1.39	4.99	10.51	32.84	3.76	13.50	28.44	88.87
ns1607	0.23	1.77	2.81	19.45	0.56	4.31	6.85	47.38
HAR20	0	6.48	9.84	30.12	0	12.40	18.83	57.65
HAR21	0	0	0	0	0	0	0	0
HAR22	0	0	0.77	3.63	0	0	1.09	5.12
HAR23	0	0.08	1.27	23.97	0	0.13	2.07	39.12
HAR24	0.10	4.02	6.01	43.50	0.21	8.63	12.91	93.44

Appendix 4 (Tables A9, A10, and A11)

Table A.9 Distance traveled, number of GPS positioning points, and estimated total cost during autumn migration of Oriental honey-buzzards (only in Japan)

Bird ID	Distance traveled (km)	Number of GPS positioning points	Estimated total cost ($\times 10^6$ kJ)
ngsku1601	1786	370	7.28
ngsku1605	1880	379	5.52
ngsku1701	1648	292	6.07
ngsku1704	1666	418	6.38
ngsku1706	1661	301	4.39
ngsku1707	1681	257	6.82
ngsku1709	1524	254	4.95
ns1621	1678	241	5.11

Table A.10 Potential number of wind turbines and farms and increased distance to circumvent them during autumn migration of Oriental honey-buzzards (only in Japan)

	Number of buffers around wind turbines and wind farms on a migration route				Increase of distance traveled assuming a bird circumvents buffers (km)			
	Buffer size (radius)				Buffer size (radius)			
Bird ID	100 m	500 m	1000 m	2000 m	100 m	500 m	1000 m	2000 m
ngsku1601	0	0	2	10	0	0	0.21	2.54
ngsku1605	0	0	2	3	0	0	0.61	2.06
ngsku1701	1	1	1	2	0	0.37	0.94	2.48
ngsku1704	0	0	0	4	0	0	0	0.92
ngsku1706	2	2	2	4	0.02	0.83	5.30	7.08
ngsku1707	0	0	3	9	0	0	0.64	10.28
ngsku1709	0	3	3	4	0	0.08	1.46	4.17
ns1621	0	0	1	2	0	0	0.12	1.47

Table A.11 Potential costs to circumvent wind turbines and farms during autumn migration of Oriental honey-buzzards (only in Japan)

	Increase of time traveled assuming a bird circumvents buffers (minute)				Increase of energetic cost to travel during additional duration (kJ)			
	Buffer size (radius)				Buffer size (radius)			
Bird ID	100 m	500 m	1000 m	2000 m	100 m	500 m	1000 m	2000 m
ngsku1601	0	0	0.27	4.80	0	0	0.79	14.08
ngsku1605	0	0	1.21	4.56	0	0	2.37	8.92
ngsku1701	0	1.13	2.89	8.69	0	2.90	7.40	22.26
ngsku1704	0	0	0	2.21	0	0	0	5.99
ngsku1706	0.04	1.92	9.81	16.26	0.07	3.35	17.13	28.39
ngsku1707	0	0	0.99	16.27	0	0	2.89	47.54
ngsku1709	0	0.54	5.71	16.44	0	1.23	13.02	37.48
ns1621	0	0	0.63	8.82	0	0	1.30	18.20

References

1. Masden, E.A., Haydon, D.T., Fox, A.D., Furness, R.W., Bullman, R., Desholm, M.: Barriers to movement: impacts of wind farms on migrating birds. ICES J. Mar. Sci. **66**, 746–753 (2009)
2. Plonczkier, P., Simms, I.C.: Radar monitoring of migrating pinkfooted geese: behavioural responses to offshore wind farm development. J. Appl. Ecol. **49**, 1187–1194 (2012)
3. Cabrera-Cruz, S.A., Villegas-Patraca, R.: Response of migrating raptors to an increasing number of wind farms. J. Appl. Ecol. **53**, 1667–1675 (2016)
4. Masden, E.A., Haydon, D.T., Fox, A.D., Furness, R.W.: Barriers to movement: modelling energetic costs of avoiding marine wind farms amongst breeding seabirds. Mar. Pollut. Bull. **60**, 1085–1091 (2010)
5. Wilmers, C.C., Nickel, B., Bryce, C.M., Smith, J.A., Wheat, R.E., Yovovich, V.: The golden age of bio-logging: how animal-borne sensors are advancing the frontiers of ecology. Ecology. **96**, 1741–1753 (2015)
6. Bridge, E.S., Thorup, K., Bowlin, M.S., Chilson, P.B., Diehl, R.H., Fléron, R.W., ... Wikelski, M.: Technology on the move: recent and forthcoming innovations for tracking migratory birds. Bioscience **61**, 689–698 (2011)
7. Watanabe, Y., Koshiyama, Y., Senzaki, H., Iseki, F.: The Field Guide to the Raptors of Japan vol. 04 Eastern Buzzard. Nakano Collotype, Okayama (2017). (in Japanese)
8. Higuchi, H., Shiu, H.J., Nakamura, H., Uematsu, A., Kuno, K., Saeki, M., Hotta, H., Tokita, K.-I., Moriya, E., Morishita, E., Tamura, M.: Migration of Honey-buzzards *Pernis apivorus* based on satellite tracking. Ornithol. Sci. **4**, 109–115 (2005)
9. Garvin, J.C., Jennelle, C.S., Drake, D., Grodsky, S.M.: Response of raptors to a windfarm. J. Appl. Ecol. **48**, 199–209 (2011)
10. Takeoka, H., Mukai, M.: Wind power generation and natural environment. Comparison of flight path of migratory birds before and after construction of wind power plants by means of seodorites. [in Japanese]. Wind Energy. **28**, 18–22 (2004)
11. Pennycuick, C.J.: Modelling the Flying Bird, vol. 5. Elsevier, London (2008)
12. Pennycuick, C.J., Åkesson, S., Hedenström, A.: Air speeds of migrating birds observed by ornithodolite and compared with predictions from flight theory. J. R. Soc. Interf. **10**, 20130419 (2013)
13. Ryan, P.G., Petersen, S.L., Peters, G., Grémillet, D.: GPS tracking a marine predator: the effects of precision, resolution and sampling rate on foraging tracks of African Penguins. Mar. Biol. **145**, 215–223 (2004)
14. Bishop, C.M., Spivey, R.J., Hawkes, L.A., Batbayar, N., Chua, B., Frappell, P.B., Milsom, W.K., Natsagdorj, T., Newman, S.H., Scott, G.R., Takekawa, J.Y., Wikelski, M., Butler, P.J.: The roller coaster flight strategy of bar-headed geese conserves energy during Himalayan migrations. Science. **347**, 250–254 (2015)

A Pioneer in Transition: Horizon Scanning of Emerging Issues in Germany's Sustainable Wind Energy Development

Johann Köppel, Juliane Biehl, Volker Wachendörfer, and Alexander Bittner

Abstract With both the United Nations Framework Convention on Climate Change (UNFCCC) and the Convention on Biological Diversity (CBD) adopted at the Rio Earth Summit in 1992, certainly no-one anticipated the challenging trade-offs between renewable energy development and conservation of biological diversity. Densely populated Germany ranks third in worldwide installed wind energy capacity, only outpaced by China and the USA so far. However, power and interest shifts via well-organised civil and political opposition indicate that efforts to reconcile climate protection and wildlife conservation cannot be taken for granted.

Funded by the German Federal Environmental Foundation (DBU), the horizon scan aimed at identifying the emerging need for collaborative action, based on the viewpoints of various stakeholders and the state of research in wildlife conservation and wind energy development. We applied a multi-faceted, inclusive, and peer-reviewed research process, building on ca. 50 explorative expert interviews, previous research, and a literature review. Interviewees ranged across academia, agencies, wind developers, consultants, associations, and environmental groups. The process yielded 18 emerging issues at the nexus of wind and wildlife, planning and technologies, and social aspects to cope with the challenges ahead. We present a majority of the issues in this chapter and conclude with guiding follow-up principles.

Keywords Horizon scanning approach · Transformation · Sustainable Development Goals · Cumulative effects · Landscape-scale conservation · Knowledge management · Planning approaches · Adaptive management

---

J. Köppel (✉) · J. Biehl
Environmental Assessment and Planning Research Group, Berlin Institute of Technology (TU Berlin), Berlin, Germany
e-mail: johann.koeppel@tu-berlin.de; juliane.biehl@tu-berlin.de

V. Wachendörfer · A. Bittner
Deutsche Bundesstiftung Umwelt, Osnabrück, Germany
e-mail: v.wachendoerfer@dbu.de; a.bittner@dbu.de

R. Bispo et al. (eds.), *Wind Energy and Wildlife Impacts*,
https://doi.org/10.1007/978-3-030-05520-2_5

1 Introduction

1.1 Framing and Motivation

With the 1992 Rio Earth Summit both the Biodiversity Convention [1] and the Climate Change Convention [2], two paramount agreements, were initiated. Certainly, the effort it would take to achieve both aspirations could not have been foreseen 25 years ago and is still pending in the twenty-first century. The contemporary 17 United Nations Sustainable Development Goals (SDGs) set tangible global goals and targets, including 'affordable and clean energy' (SDG 7) and 'climate action' (SDG 13) [3]. All parties to the resolution need to 'increase substantially the share of renewable energy in the global energy mix by 2030' (goal 7.2 Agenda 2030) and to 'integrate climate change measures into national policies, strategies and planning' (goal 13.2 Agenda 2030).

In analogy to the SGDs, Rockström et al.'s 'planetary boundaries' [4, 5] focus on sustainability challenges, for instance, biosphere integrity (i.e. biodiversity and functionality of ecosystems), climate, and land use change. The concept identifies 'tipping points' (e.g. melting of the Greenland ice sheet or the decay of coral reefs), whose trespassing could shift the entire earth system into instability.

Staying in a 'safe operative space' (i.e. within the planetary boundaries) offers the opportunity for development in a sustainable way whilst maintaining the biosphere integrity as an essential precondition [4, 5]. Folke et al. [6] stated that the SDGs 'life on land' (SDG 15), 'life below water' (SDG 14), 'clean water and sanitation' (SDG 6), and 'climate action' (SDG 13) are providing a basis for biosphere integrity and for all the other 13 SDGs, which belong to the societal and economic sectors (Fig. 1).

The German *Energiewende* contributes significantly to extending renewable energy and to staying within a safe operative space. However, designing the transformation of the energy sector as a complex system proved not to be an easy assignment, since many of the agents participating in these processes are used to thinking and acting in rather sectoral ways. In contrast, it is necessary to refer to regional landscapes and the relations between agents and subagents in 'socioecological systems' (SES).

Referring to the planetary boundaries and to the SDGs, the German Federal Environmental Foundation (DBU) updated their guidelines in 2016, including new concepts and environmentally sound solutions for renewable energy. Against this background, the DBU decided to launch a horizon scan, conducted by Berlin Institute of Technology's (TU Berlin's) co-authors of this chapter. At the nexus of wildlife conservation, planning and technologies, and social issues (e.g. fair allocation processes), the horizon scan involved stakeholders from academia, agencies, the wind energy sector, consultants, and nongovernmental organisations (NGOs). Previously, horizon scans were elaborated for a variety of issues, such as nature conservation and biological diversity, with the latter annually undertaken by Sutherland et al. [7–12].

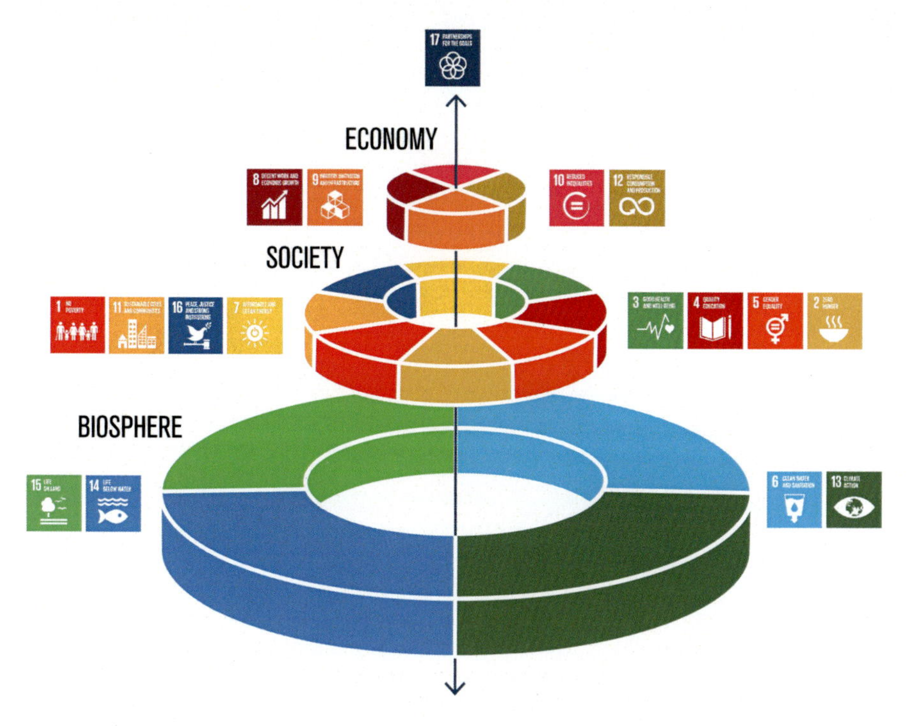

Fig. 1 Combination of the SDGs with the planetary boundaries concept resulting in the so-called wedding cake with biosphere integrity and the adjacent SDGs as base for sustainability and resilience. (Copyright: Azote Images for Stockholm Resilience Centre)

1.2 The Challenged Pioneer in Transition

In 2016, renewable energies contributed 29.5% to the German electric power production (of which land-based wind energy amounted to 10.3% and offshore 2%) [13]. With respect to cumulative installed wind energy capacity, Germany ranks third after the People's Republic of China (168 GW) and the United States of America (82 GW), totalling 50 GW in 2016 [14].

Yet, considering the installed capacity per land area, the relative density in Germany (or the CWW 2017 host Portugal) outpaces the ones in the USA or China (Fig. 2). The challenges at hand even increase when considering the actual suitability of land for siting wind energy facilities (e.g. due to wind conditions, distance to settlements, wildlife, among other relevant planning concerns).

Whilst the German pioneering phase started as early as in the mid-1970s, the provision of subsidies under the Electricity Feed-in Act 1991 paved the way for wind energy operators to enter the market in the early 1990s [15]. With the Renewable Energy Sources Act (EEG) in 2000 and its amendments in 2004 and 2009, feed-in tariffs for land-based wind turbines were reduced, but renewable electricity was given feed-in priority into the transmission grid over

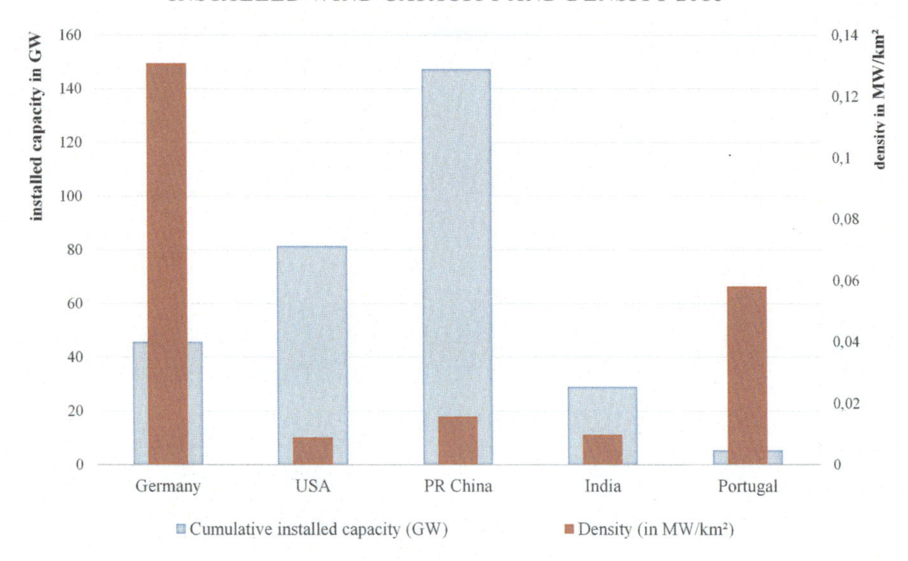

Fig. 2 Installed wind capacity and density in selected countries, as of 2016. (Data sources: Cumulative installed capacity – Global Wind Energy Council 2016, p. 12; Land area – Statistisches Bundesamt, Destatis, https://www.destatis.de/EN/FactsFigures/CountriesRegions/InternationalStatistics/Country/Country.html)

conventional sources. The loss of public acceptance for nuclear energy and the subsequent decision to phase out nuclear power entirely by 2022 are other drivers for the German energy transition [16]. Yet, Germany as one of the *Energiewende* pioneers is undergoing substantial transitions, since the latest EEG amendment 2017 introduced a market-oriented auctioning system for renewable energy projects. At the same time, the EEG 2017 set trajectories, which targeted the installation of land-based wind energy in 2017–2019 at annually 2,800 MW and in 2020 at 2,900 MW. Offshore trajectories were set to 6,500 MW by 2020 and 15,000 MW by 2030 [17].

Yet, the SDGs equally comprise conservation goals of terrestrial and marine ecosystems – 'life below water' (SDG 14) and 'life on land' (SGD 15) [3]. Achieving both, biodiversity and climate protection goals, highlights the necessity to engage in balancing so-called 'green vs. green' dilemmas [cf. 18] and to overcome unintended trade-offs between SDGs. Whilst Germany has installed more than 28,000 wind turbines [19], it has become increasingly difficult to further make the *Energiewende* a success. Inter alia, concerns address assumed decreases in property values [20–22] or fuel equity debates on sharing revenues from wind energy projects. Opposition groups engage in planning processes at various governance tiers and campaign against further wind energy development. Yet, whilst local opposition addresses, for example, landscape scenery impacts [23], opponents tend not always to reveal such latent motives [24] but pinpoint legally stronger enforced wildlife concerns [25].

Emerging issues in the German wind energy and wildlife arena apply as well on an international scale, since eventually other countries will face similar challenges when developing wind energy. Thus, the horizon scan aimed at identifying emerging issues for a sustainable wind energy development and applied a multi-faceted, inclusive, and peer-reviewed research process.

2 Methods

2.1 Horizon Scanning

Horizon scanning is – according to the European Commission – a technique 'for detecting early signs of potentially important developments' [26]. We pursued a 'manual-combined approach' [27] to horizon scans by using expert interviews, surveys (using SurveyMonkey), expert workshops and a final symposium, as well as a targeted literature review and by attending conferences and seminars (inter alia NWCC 2016, CWW 2017) (Fig. 3). Thus, we adopted an 'issue-centred' scanning approach [27], starting from a wider range of predefined emerging issues to either verify these issues, identify modifications, or reveal new topics [27].

Our research was initiated from 7 preliminary meta-topics (population modelling and impact assessment, deterrence, repowering and planning, cumulative impacts, adaptive management, social impact assessment, and sustainability appraisal) based on prior research at TU Berlin's Environmental Assessment and Planning Research

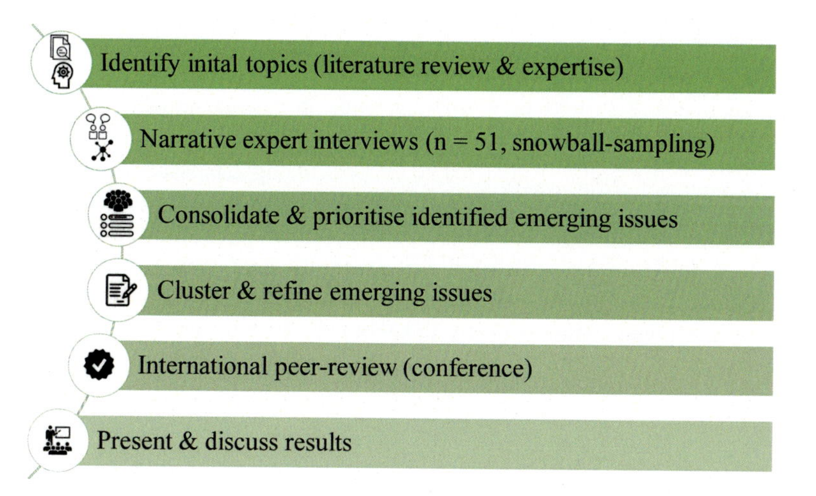

Fig. 3 Workflow and process steps of horizon scanning for emerging issues in sustainable wind energy development

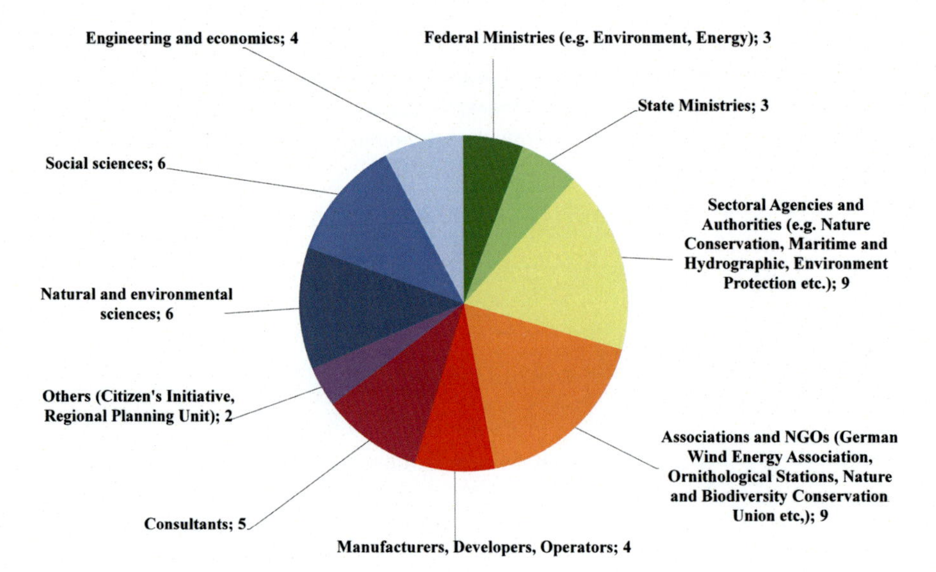

Fig. 4 Number of expert interviews by stakeholder groups ($n = 51$)

Group [e.g. 15, 18, 28–33] and relevant funding experience of the DBU (https://www.dbu.de/2535.html).

Over the course of 18 months, the research team conducted 51 semi-structured, narrative expert interviews with key stakeholders from academia, agencies, wind energy sector, consultants, and the civil society (Fig. 4). Experts were selected by snowball sampling – a convenience sampling technique allowing researchers to ask participants for additional peers [34, 35]. Jointly, the horizon scan assisted in identifying 18 nascent issues at the nexus of wind and wildlife, planning and technologies, and social aspects.

Several involved experts ($n = 24$) convened in a workshop in April 2017 to discuss preliminary results and identified emerging issues. A supplementary online survey (19–28 April 2017) contributed in prioritising the issues. Subsequently, the research team clustered the identified emerging issues into three categories: fact-checking, model approaches, and proof of concept. These clusters indicate overarching tendencies in the need for action. A final symposium took place in September 2017, offering a forum for discussions and for drawing conclusions collaboratively with involved stakeholders. Participants and further peers were invited to comment on the publicly available draft version of the final research report (in German), thus safeguarding an inclusive and peer-reviewed process.

2.2 *Study Limitations*

The horizon scan was not necessarily a representative study, yet based on viewpoints of multiple stakeholders and the state of research. It cannot claim to be exhaustive and exclusive. Interviews with experts in the financial and banking sector were beyond the scope and might require further investigations. Specific judicial aspects were not considered either, since such limitations should not affect the genesis of visionary ideas in this horizon scanning process. Nevertheless, our research design offered various opportunities to comment and engage, to attract additional experts or stakeholder groups, and to add further emerging issues to the agenda.

The horizon scan predominantly focused on land-based wind energy, as this approach initially strived at a decentralised *Energiewende*. Additionally, various research programmes have been conducted in recent years in the offshore realm. For the time being, an integral view of the three dimensions (wildlife conservation, planning and technologies, as well as societal aspects and participation) seemed most profitable and achievable for land-based wind energy in a German context.

A larger and more far-reaching horizon scan could be achieved by including novel scanning technologies, e.g. blogging and microblogging (Twitter and other social media) [27]. Searching through these social media platforms might facilitate obtaining new information faster and in real time [27]. In using both microblogging and Delphi-like scoring approaches, an international team identified 15 nascent issues in nature conservation in 2017, for example, covering new developments in energy storage and the offshore wind energy sector (floating wind turbines) [9].

3 Results and Discussion

Both the expert interviews and the literature review yielded a meta-structure of 18 emerging issues, spanning the 3 dimensions of wildlife conservation, planning and technologies, as well as societal aspects and participation (Fig. 5). Each emerging issue addresses various options, constituting a catalogue that might contribute either per se or favourably within integrated programmes.

Consequently, the research team clustered the nascent issues into three categories: fact-checking, model approaches, and proof of concept. These clusters indicate overarching tendencies in the need for action:

- The cluster 'fact-checking' comprises topics that require evidence-based scrutiny and empirical corroboration of commonly held assumptions.
- The cluster 'model approaches' addresses issues that require innovative and model approaches and often demand collaborative action of multiple stakeholders. Reality laboratories and other innovative solutions would come handy to address the issues.

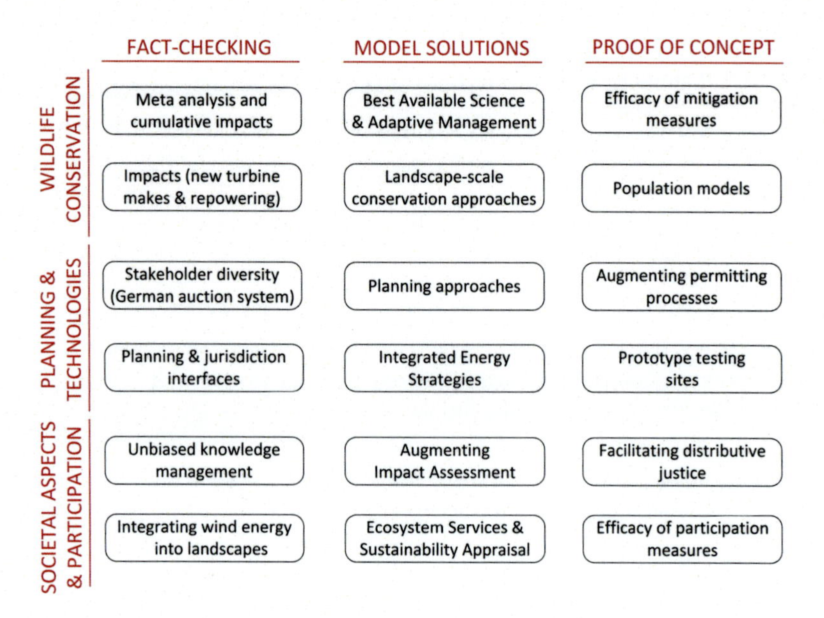

Fig. 5 The horizon scan yielded 18 emerging issues in sustainable wind energy development spanning the 3 dimensions of wildlife conservation, planning and technologies, as well as societal aspects and participation (as of December 2017)

- With the cluster 'proof of concept', we aggregated all issues that require prototyping and pilot studies to eventually test and verify hypotheses – last but not least in a German context.

In the following, 11 emerging issues address a wind and wildlife community and offer insights of relevance for an international audience. The remaining seven issues were not introduced here as they focus explicitly on the German context and revolve primarily around socio-economic aspects of wind energy. The complete report (in German) is available online at https://www.dbu.de/projekt_33315/01_db_2409.html.

3.1 *Fact-Checking*

3.1.1 Meta-analyses and Cumulative Impacts

The possibility to provide added value in aggregating data and conducting meta-analyses was highlighted [36, 37] – notably for assisting cumulative impact analyses. For calculating the net footprint of wind energy development on wildlife, for instance, a holistic overview and at best nationwide data would be required. Still questions remain whether data were gathered standardised in pre- and post-construction assessments and whether further data (weather, land cover, etc.)

are available [38], e.g. to derive cumulative, indirect, and multivariate impacts. Our findings are in line with the prior discourse on aggregating pre- and post-construction data (e.g. from ornithological assessments) and monitoring results [39, 40]. The German ornithological stations have been collecting information on bird and bat collision at wind farms [41]; yet, working with a data set comprised of both random sampling and standardised fatality searches. Further experts voiced concerns over the feasibility of these meta-analyses for land-based wind energy impacts due to the heterogeneity of available data sets, notwithstanding relevant guidance documents.

In contrast, a more standardised approach was adopted offshore: Developers are obliged to conduct standardised monitoring during the construction of all offshore wind turbines, whilst monitoring during the operational phase can be an additional requirement from the permitting authority (e.g. via substantiating permit requirements) [42, 43]. The transboundary analysis of data could prove beneficial – especially with respect to cumulative impacts pertaining to relevant offshore wind farm clusters. Experts addressed the lack of harmonised approaches in conducting such analyses across borders.

3.1.2 Planning and Jurisdiction Interfaces

The establishment of exchange interfaces between planners and jurisdiction has been identified as another notable remit. Litigation in Germany has risen substantially over the years, with new and often abstract or even vague legal terms arising from settled case law, e.g. the so-called 'soft no-go areas' for wind energy [44] or an approach to a 'significant increase in collision risk' for wildlife [45]. Critique has been voiced that jurisdictions were not always considering planning feasibility [46], often leaving practitioners and (regional) planners in the dark to find appropriate indicators in operationalising new rulings. With the goal of uncovering efficient interfaces, a planning and jurisdiction platform was proposed to regularly engage and sensitise judges for a planner's expertise.

3.1.3 Unbiased Knowledge Management

Interviewees revealed they struggle with suspect of partiality and mistrust among stakeholders, e.g. with respect to the contracting of consultants. Often, neither wildlife agencies nor environmental NGOs felt comfortable with consultants appointed by wind developers, presuming vested interests. In turn, wind developers were not feeling content either once they encountered agency officials, who were members of opposing environmental NGOs at the same time. Collaborative boards – i.e. a jointly established group of people both from the wind energy sector and wildlife conservation arena – might assist in appointing the experts and consultants together. Closer co-operation – early on in the wind energy project design and planning phase – might increase transparency and reveal hidden resentments [cf. 47]. Additionally, such collaborative boards might also assist in reviewing interim and major findings – at least in disputed or sensitive cases, as, e.g. Canadian law offers for the application of review panels in certain environmental assessment processes on federal and provincial levels [48].

With respect to the management and dissemination of knowledge, various interest groups engage in diffusing study insights, by disseminating spreadsheets and 'fact' sheets, for instance. However, such 'third-party' communication raises some scepticism when diffusing meta-information derived from original scientific findings, particularly with respect to limitations of scientific findings and the transferability of results. Findings by Wolters et al. [47] confirm that 'the scientist's ability to communicate effectively with people both within and outside of the scientific field is important'. The researchers themselves are most qualified to reach out and communicate their insights, case-specific limitations, and transferability of their findings. Consequently, relevant outreach work packages need to be embedded within research and development projects, e.g. in supporting communication and dissemination – if necessary with plain language summaries and writing toolboxes [49].

3.1.4 Integrating Wind Energy into Landscapes

Landscape Narratives The question arose if and when wind turbines will ever be perceived as constituting elements of our cultural landscapes [cf. 50]. This is closely linked to various questions: How do residents perceive landscapes before and after the construction of wind turbines? Which societal framings (familiarisation patterns, aspirations, experience, and recollection) can be identified, for both residents and visitors? Do the actual nuisances for residents prove to be as strong as previously envisaged in the permitting phase? These issues might matter substantially for the repowering of wind farms.

Albeit studies with relevant mid- or long-term perspectives are missing, representative, annual, and recurring opinion polls [e.g. 51–53] give first insights and may detect trends. Nonetheless, a study project from a TU Berlin master's course [54] offered a methodological approximation to the topic. Based on a baseline survey from 2005, students conducted a follow-up survey in 2016: The working hypothesis – that the residents' acceptance decreases with increasing numbers of wind farms in their neighbourhood – was rejected. The study further indicated that younger respondents seemed more accustomed to wind farms and showed higher acceptance levels than senior generations [54].

Further research might address this in more depth, aiming to identify relevant 'tipping points' (i.e. a moment of pivotal change) of residents' perceptions. These insights might assist in understanding whether wind turbines might eventually constitute an integral part of our cultural landscapes.

Citizens' Opposition Against Wind Energy Development Another emerging issue pertains to the question of citizens' initiatives and wind energy opponents. Local citizens' groups in Germany are usually well-organised in regional, supra-regional, national [55], and even European [56] umbrella organisations. The latter provide leaflets, scripted arguments against wind energy, and a professionalised network of experts from various disciplines (medical doctors, ornithologists, engineers, lawyers, politicians, etc.). Recent studies of citizens' initiatives against the transmission grid expansion reflect the broad political and social spectrum of such groups [57]. Such findings question the 'not in my backyard' (NIMBY) concept [58–61] and whether opposition against wind energy still can be considered a local phenomenon. Nonetheless, climate change – and *Energiewende* – sceptics can be identified behind the leading and resourceful umbrella organisations [cf. 62]. It became apparent that local-level approaches are not sufficient in resolving conflicts over wind energy development. For instance, overarching mechanisms (e.g. the Renewable Energy Sources Act, EEG, or the privileged status for wind development in peripheral areas according to the Federal Building Code, BauGB) became subject to protest from a German umbrella initiative against wind energy [63], thus calling into question the energy system's transformation (*Energiewende*) [64]. Hence, a genuine and all-encompassing societal discourse on climate change and the energy transition would still be required [65], calling for a decent arena besides the wind-wildlife discourse.

3.2 Model Approaches

3.2.1 Best Available Science and Adaptive Management

Best Available Science (BAS) Mandate A set of regulations governing environmental conservation and management in the USA stipulate that best available science (BAS) be used as the basis for policy- and decision-making [66]. This BAS mandate aims at achieving high-quality scientific results, defined by Sullivan et al. [66] in, e.g. applying standardised methods for gathering data and clearly documenting the research process, safeguarding statistical rigour and concise logic, and enabling peer review of scientific insights. Consequently, a study's limitations and the transferability of results should be clearly stated [47]. The mandate requires also to differentiate to which degree results and conclusions are based on expert judgement, on the one hand, or empirical evidence on the other hand. A similar BAS mandate, as it is applied, e.g. in the context of the Endangered Species Act (ESA) in the USA [67], has not been established in Germany.

For example, it is of major interest to which degree species-specific buffer zones to wind energy facilities around breeding and resting sites [68] are based on empirical research results. Expert judgement is neither an inferior approach per se nor are expert opinions and prerogatives disregarded by the courts. German settled case-law ruling allows thus far for expert prerogatives in cases, where neither methods nor scientific evidence have been established [69]. Yet, a sound BAS mandate also in Germany would presumably serve well to inform the decision-making processes better.

Adaptive Management (AM) in a German Context Adaptive management [70] contributes as well to the BAS mandate, in pinpointing the uncertainties in, inter alia, collision risk estimation, displacement risk assessments, and the success of mitigation measures [71].

Uncertainties pertaining to both the actual impacts of wind energy on wildlife and mitigation measures remain [72]. For instance, natural dynamics (e.g. fluctuating populations or alternating roosting sites of birds of prey) have received little attention in prior planning and approval procedures [72]. In contrast, the current legal framework seems to challenge experts with respect to the question whether additional mortality would significantly influence the (local) population's well-being [40, 73, 74]. In building on ongoing learning and adaptation processes, AM provides means to recognise and overcome uncertainties [70, 75, 76]. Yet, permitting authorities would require supplementary capacities to supervise and effectively test AM. The major question revolves around judicial aspects and the degree of permits' flexibility. To contrive a dynamic (i.e. conditional but valid) permit does not allow for simply trial-and-error approaches [72].

Model projects might be of assistance in exploring AM in practice. Ideally, all relevant stakeholders should engage in collaborative action, trying to create mutual trust and at best win-win situations for all parties concerned [72]. Sound AM approaches should not generate burdens for one party only (e.g. intensified curtailment regimens) but also consider scenarios with benefits for wind farm operators (in case of alleviating monitoring results) [71].

3.2.2 Landscape-Scale Conservation Approaches

Recent research indicated a possible risk for a significant population decline for red kite (*Milvus milvus*) [40, 77] or even common buzzard (*Buteo buteo*) [38, 77], raising the question whether further wind energy development needs to be restricted. To achieve both biodiversity and climate protection targets, supplementary approaches might assist so far established mitigation measures at wind farm/turbine/project level. Establishing conservation approaches at larger scales, e.g. at a landscape level, was identified as a relevant topic over the horizon scan discourses. Exploring concepts to stabilise populations at a supralocal scale might prove to be of value in case the local mitigation sequence and site-specific measures cannot provide remedy alone. In addition, cumulative impacts might be addressed at the supralocal level by means of landscape-scale approaches. This has been conducted for the Indiana bat (*Myotis sodalis*) and the bald eagle (*Haliaeetus leucocephalus*) in the US-Midwestern Wind Energy Multi-Species Habitat Conservation Plan (HCP) [78]. In another effort, the NWCC Sage-Grouse Research Collaborative was established [79] – funded from agencies, the wind industry, and NGOs.

These examples illustrate that a change in perspective is required, centring approaches on the target species in question and working collectively with all land users, who possibly influence the target species' well-being. Such model projects require an unbiased broker, working collaboratively with contributing land users (e.g. agricultural sector, forestry, transmission grid operators, renewable energy industry). One feasible model to sustain efforts financially could be a common fund pooling financial resources from all developers in the region (and – at best – further funds from third parties).

Another issue in this context highlighted the need to address (regionally or – in some cases – statewide) increasing populations of endangered species. Such species, e.g. the Eurasian eagle-owl (*Bubo bubo*) in Germany [80], hold the same (strict) legal protection status as species with seriously declining populations. In the USA, population trajectories and a potential biological removal (PBR) model [81–83] were applied for the bald eagle (*Haliaeetus leucocephalus*). In estimating the 'number of individuals that could be killed before a population will fall below a size considered sustainable' [cf. 83], PBR is a 'reference point' approach to assessing and setting limits on human-induced mortality to species [84]. At the same time, O'Brien et al. [85] raise critique on PBR models as they rely 'on a number of implicit assumptions, particularly around density dependence and population trajectory that limit its applicability in many situations'.

3.2.3 Planning Approaches

Germany has an ambitious planning and zoning system that serves in governing the spatial development of wind energy. It follows a decentralised approach with several planning levels [86]. The states can set renewable development trajectories [87], and regional planning entities contribute a vital link between the state's objectives and local perspectives. Planning regions designate priority areas for wind development [86], whilst the municipalities can further zone the actual sites (concentration zones) [88]. Hot spots of protected wildlife, recreational areas, distances to settlements, and heritage sites among others are considered early in the planning processes [89].

As an outcome, concentration zones have been identified, to counteract an all-too-scattered wind energy development [88, 89]. However, in some regions this led to critical concentrations and stigmatic 'energy landscapes', i.e. visually dominant energy facilities, raising relevant public concerns [90]. Thus, planners seemed no longer convinced that the concentration paradigm for wind energy constitutes a one-and-only approach. Alternative models need to be explored and/or prototyped in scenario analyses. The puzzling question of a 'least impact wind farm' siting concept was highlighted as an emerging issue, rethinking the established planning domain with exploring legal ramifications of other approaches, too.

3.2.4 Ecosystem Services and Sustainability Appraisals

Sustainability appraisals might provide a means to balance the 'net footprint' of wind energy development and account for debits and credits at the same time [91, 92]. However, German permitting processes so far consider primarily adverse impacts pertaining to wind farms. Downsides are debited, whilst beneficial ecosystem services are not credited yet. Manufacturers, developers, and operators therefore voiced an interest in addressing this 'blind spot' of permitting processes and crediting the benefits of wind energy (e.g. decarbonisation) as well. Scholarly work emphasises an integrated ecosystem services (ESS) approach [91–93], identifying debits and credits for wind farms. Yet, these approaches bear certain risks of oversimplification and still require a more in-depth assessment of ESS. Nevertheless, sustainability appraisals assist in transparently communicating not only the inherent trade-offs but also co-benefits of renewable energy (e.g. job and regional added value creation) [94, 95]. This emerging issue still needs to be discussed thoroughly and further grounded. It could serve as well to explore cross-sectoral sustainability appraisals, for example, via interlinking the German Energiewende and the proclaimed Verkehrswende (transport and mobility transition).

3.3 *Proof of Concept*

3.3.1 Efficacy of Mitigation Measures

Efficacy of Measures to Avoid, Minimise, and Reduce Impacts In Germany, there is an ongoing controversy whether only careful planning and siting measures constitute proper mitigation approaches (i.e. avoiding according to May [96], e.g. by establishing buffer zones around breeding sites) or to which degree more detailed measures matter as well (e.g. land use management to luring birds away to more attractive foraging sites near-by). Much is governed in Germany by mostly statewide guidance [97], which sets up standards for baseline surveys and relevant methods but also recommends sequential mitigation measures (avoid, minimise, reduce, cf. May [96]). Nevertheless, no proof of concept, i.e. the mitigation measures' actual efficacy, has been widely established yet [31, 32, 96]. In turn, interviewed wind energy developers and operators stated that they must be able to rely on the feasibility of recommended mitigation approaches in relevant guidance. At the end of the day, proof for efficacy of mitigation measures has been strongly longed for in the horizon scan from different stakeholder groups, from both conservation and industry sides. In turn, compensatory mitigation has been given a second-rate treatment in contemporary German practice in the wind energy context. This would require a European Protected Species derogation licence (comparable to an incidental take permit in the USA). To date, such derogation licences are not regularly practised, since proponents have to justify the non-existence of reasonable alternatives (mostly other locations) and proof that the conservation status of the population of a species will not be deteriorated. In contrast, impacts on landscape scenery are considered unavoidable per se and require in-lieu fees for any compensatory measures [98].

Cost-Effectiveness This emerging issue pinpoints the question of maximum operating restrictions, which might still be tolerable for operators. Therefore, the cost-effectiveness of mitigation measures might be of relevance as well. In Germany, curtailment is required not only for bat species but also for specific bird events, for example, during and after mowing to minimise collisions of raptors. Additional limits for human health protection set standards, e.g. on the maximum exposure to the periodical shadowing effect of wind turbines (annually max. 30 h, daily max. 30 min [99]). The cost-effectiveness of various curtailment measures (bat algorithms, emission control standards, etc.) would be of interest beyond their wildlife and health benefits. The Water Framework Directive, respectively, requires member states to 'make judgements about the most cost-effective combination of measures' [100] in an economic analysis; possibly a similar approach might be explored for wind energy as well.

Another emerging issue addressed 'retrofit' mitigation measures for approved and operating wind farms. For instance, the relocation of less ideally sited turbines (e.g. high fatality rates on-site) might matter in specific cases and be supported by using abound in-lieu fee money. In Germany in-lieu fees often need to be paid by developers for the lack of effective landscape scenery compensation [98].

Deterrence Albeit recent studies on the efficacy of deterrence measures are underway (e.g. DTBird in Norway [101], Switzerland [102], Sweden [103], and the USA [104, 105]; Frontier Wind [106]; RNRG deterrent [107]; IdentiFlight [104]), a proof of concept for Germany is pending. One of the studies on the efficacy of deterrence measures (Department of Energy, DoE-funded) aims at comparing the reduction levels of an ultrasonic acoustic (RNRG) deterrent with operational minimisation (i.e. curtailment by feathering blades to 5.0 m/s or 2.0 m/s above the pre-set operating conditions) [107]. This approach addresses the combination of both mitigation measures to test 'whether there is an additive effect, furthering the reduction in bat fatalities' [107]. Another DoE-funded project evaluates a camera-based eagle detection system (IdentiFlight deterrent). It aims at determining how this technology compares to biological observers in identifying and in reducing collision risks for eagles [104]. Yet, research results and deterrent technologies might not easily be transferable to other settings, e.g. due to the on average higher wind turbines in Germany and settings in various land covers and rugged terrains (e.g. forests, forested midlands). Contemporary settled case-law rulings refer to a missing proof of concept and have denied relevant permits so far [108]. Yet, an automated bird detection system will be tested in Southern Germany as part of a recent wind energy research cluster [109].

3.3.2 Population Models

In the USA, population models have been widely applied by wildlife agencies (US FWS), e.g. for conceptualising eagle management [82]. Lately, matrix models were set up in Germany for some raptor species [77]. This recent German matrix modelling approach [cf. 38] stressed the need for further elaboration of population models in conservation, i.e. to support the modelling approaches' feasibility, accountability, and limits. To inform modelling, the relevant data base needs to be improved substantially, and sound Before-After-Control-Impact (BACI) studies would be required whilst focusing on the actual comparability of pre- and post-construction studies as well [110]. This emerging issue is also linked to the cost-efficiency topic, since advanced and contemporary telemetry technologies [e.g. 111] might prove less resource-intensive compared to customary methods. Interviewees mentioned that a comparative study could collate wind concentration zones with reference zones with fewer installed turbines. Without advancements in population modelling, the still challenging issue of identifying significant mortality thresholds in permitting procedures might hardly become better elaborated.

3.3.3 Efficacy of Participatory Approaches

The horizon scan revealed interest in measuring the efficacy of mitigation measures not only in wildlife conservation. Inclusive and good-practice participatory approaches were addressed as well. Developers have a legitimate interest in learning whether extensive participation in the planning process or financial participation models can actually foster local wind farm acceptance. Furthermore, radar-based on-demand turbine lighting technologies for approaching aircrafts can avoid permanent night-time lighting near settlements. Since they are rather costly, their actual efficacy in increasing residents' well-being was of interest in the horizon scan as well.

4 Conclusions

To sum up, the window of opportunity is very narrow to cap the global CO_2 emissions, and we need to decide now about the transformation of the energy sector, notwithstanding challenging dilemmas such as climate protection vs. biodiversity conservation or 'green vs. green'. Consequently, the planetary boundaries *and* the SDGs need to be addressed in co-operation and at best collaboration with all relevant agents, building regional stewardships [6, 112, 113]. At the end of the day, local and regional actions should result in significant climate mitigation effects even on a higher scale (national, global). Our horizon scan indicates that only inter- and transdisciplinary approaches will have the potential to facilitate a sustainable energy transition within this window of opportunity. From our point of view, common, sector-bridging concepts from the following criteria (Table 1) can lead to overcoming the recent dilemmas associated with the transformation of our still-too-fossil-driven energy systems. Moreover, as far as the supportive CWW series is concerned, opening up to involved societal topics might help to further paving the road towards a sustainable wind energy development worldwide.

Table 1 Criteria for co-operative research and development projects, derived from the horizon scan

Criteria for co-operative research and development projects	
Collaborative action	Collaborative action of wildlife conservationists and climate actors (e.g. the renewable energy sector) requires creating trust and *win-win* situations for all stakeholders involved. This allows experts from various disciplines (both from research and practice) to better integrate knowledge from different disciplines [114] and engage on a regular basis. An unbiased discussion forum should be established by collaborative approaches, assisting in generating and diffusing comprehensible and credible results [115]. Advantages include the opportunity to leverage funds (e.g. contribution of nonmonetary resources, technology, and copayments) [116], access to (interim) research results, and fostering transparency [47, 117]
Perception of Sustainable Development Goals and planetary boundaries	To address the different dimensions (wildlife conservation, planning and technologies, and societal issues), research and development projects should address the SDGs and planetary boundaries candidly. Conflicting goals should be well considered and reflected, for example, in case wind energy was to be precluded from forest habitats, which other renewable energy could be developed in the same region?
Early on solution orientation	Albeit a project might intend to explore basic empirical findings or set up ambitious modelling efforts, for example, conceivable solutions should be envisaged from the very beginning. To sum up, research foci should be solution-oriented early on and can be further adjusted over the research and development process
Cost-benefit effectiveness	Taking into account cost-benefit considerations for mitigation measures might – albeit their benefits for wildlife or health protection – prove relevant, as well. The Water Framework Directive, for example, requires member states to 'make judgements about the most cost-effective combination of measures' [100]
Best available science (BAS) mandate	To increase credibility and safeguard quality in research and development as well as good practice, a thorough application of the BAS mandate is paramount. Proponents, responsible agencies, and experts involved must commit to reproducibility and data availability, include peer-reviewed processes, and indicate exactly whether specific findings are evidence-based or primarily reflect expert judgement
Communication	Recent experiences with false assertions (so-called alternative facts) in a post-truth era shows that the fine line between false assertion and facts is often hard to illustrate [cf. 118, 119]. The altered use of media often assists in circulating commonly held assertions [116]. This demonstrates the need for clearly communicated limitations and transferability of studies and results [64]. Often, this can only be adequately done by the authors of the study themselves, pinpointing the potential of integrated communication modules in research projects

Source Attributions

Icons by Flaticon Basic Licence CC3.0 (Creative Commons) by https://flaticon.com	
Search free by Freepik	
Studying by Freepik	
Chat by Good Ware	
Network by Smashicons	
Crowd of users by Freepik	
List by Freepik	
Contract by Freepik	
Approval by Freepik	
Classroom by Freepik	

Acknowledgements The horizon scan was supported by the German Federal Environmental Foundation (Deutsche Bundesstiftung Umwelt, DBU). We particularly thank our interview partners, panellists, and renowned experts for their contributions and valuable comments on earlier versions of the horizon scan study. We want to express our gratitude towards our two reviewers, whom we are highly indebted to for their valuable feedback.

References

1. United Nations (UN): United Nations Convention on Biological Diversity (CBD) (1992)
2. United Nations (UN): United Nations Framework Convention on Climate Change (UNFCCC) (1992)
3. United Nations (UN): Transforming our World: the 2030 Agenda for Sustainable Development. A/RES/70/1 (2015)
4. Rockström, J., Steffen, W., Noone, K., Persson, Å., Chapin III, F.S., Lambin, E., Lenton, T., Scheffer, M., Folke, C., Schellnhuber, H.J., Nykvist, B., Wit, C., de Hughes, T., van der Leeuw, S., Rodhe, H., Sörlin, S., Snyder, P., Costanza, R., Svedin, U., Falkenmark, M., Karlberg, L., Corell, R., Fabry, V., Hansen, J., Walker, B., Liverman, D., Richardson, K., Crutzen, P., Foley, J.: Planetary boundaries: exploring the safe operating space for humanity. Ecol. Soc. **14**(2), 32 (2009)
5. Steffen, W., Richardson, K., Rockström, J., Cornell, S.E., Fetzer, I., Bennett, E.M., Biggs, R., Carpenter, S.R., Vries, W., de Wit, C.A., de Folke, C., Gerten, D., Heinke, J., Mace, G.M., Persson, L.M., Ramanathan, V., Reyers, B., Sörlin, S.: Sustainability. Planetary boundaries: guiding human development on a changing planet. Science (New York, N.Y.). (2015). https://doi.org/10.1126/science.1259855
6. Folke, C., Biggs, R., Norström, A.V., Reyers, B., Rockström, J.: Social-ecological resilience and biosphere-based sustainability science. E&S. (2016). https://doi.org/10.5751/ES-08748-210341
7. Sutherland, W.J., Allison, H., Aveling, R., Bainbridge, I.P., Bennun, L., Bullock, D.J., Clements, A., Crick, H.Q.P., Gibbons, D.W., Smith, S., Rands, M.R.W., Rose, P., Scharlemann, J.P.W., Warren, M.S.: Enhancing the value of horizon scanning through collaborative review. Oryx. (2012). https://doi.org/10.1017/S0030605311001724
8. Sutherland, W.J., Woodroof, H.J.: The need for environmental horizon scanning. Trends Ecol. Evol. (2009). https://doi.org/10.1016/j.tree.2009.04.008
9. Sutherland, W.J., Barnard, P., Broad, S., Clout, M., Connor, B., Côté, I.M., Dicks, L.V., Doran, H., Entwistle, A.C., Fleishman, E., Fox, M., Gaston, K.J., Gibbons, D.W., Jiang, Z., Keim, B., Lickorish, F.A., Markillie, P., Monk, K.A., Pearce-Higgins, J.W., Peck, L.S., Pretty, J., Spalding, M.D., Tonneijck, F.H., Wintle, B.C., Ockendon, N.: A 2017 horizon scan of emerging issues for global conservation and biological diversity. Trends Ecol. Evol. (2017). https://doi.org/10.1016/j.tree.2016.11.005
10. Sutherland, W.J., Aveling, R., Brooks, T.M., Clout, M., Dicks, L.V., Fellman, L., Fleishman, E., Gibbons, D.W., Keim, B., Lickorish, F., Monk, K.A., Mortimer, D., Peck, L.S., Pretty, J., Rockström, J., Rodriguez, J.P., Smith, R.K., Spalding, M.D., Tonneijck, F.H., Watkinson, A.R.: A horizon scan of global conservation issues for 2014. Trends Ecol. Evol. (2014). https://doi.org/10.1016/j.tree.2013.11.004
11. Sutherland, W.J., Fleishman, E., Mascia, M.B., Pretty, J., Rudd, M.A.: Methods for collaboratively identifying research priorities and emerging issues in science and policy. Methods Ecol. Evol. (2011). https://doi.org/10.1111/j.2041-210X.2010.00083.x
12. Sutherland, W.J., Broad, S., Caine, J., Clout, M., Dicks, L.V., Doran, H., Entwistle, A.C., Fleishman, E., Gibbons, D.W., Keim, B., LeAnstey, B., Lickorish, F.A., Markillie, P., Monk, K.A., Mortimer, D., Ockendon, N., Pearce-Higgins, J.W., Peck, L.S., Pretty, J., Rockström, J., Spalding, M.D., Tonneijck, F.H., Wintle, B.C., Wright, K.E.: A horizon scan of global conservation issues for 2016. Trends Ecol. Evol. (2016). https://doi.org/10.1016/j.tree.2015.11.007
13. Agora Energiewende: Die Energiewende im Stromsektor. Stand der Dinge 2016. Rückblick auf die wesentlichen Entwicklungen sowie Ausblick auf 2017, Berlin. https://www.agora-energiewende.de/fileadmin/Projekte/2017/Jahresauswertung_2016/Agora_Jahresauswertung-2016_WEB.pdf (2017). Accessed 16 Nov 2017
14. Global Wind Energy Council (GWEC): Global Wind Report. Annual Market Update 2016, Brussels (2016)

15. Bruns, E., Ohlhorst, D., Wenzel, B., Köppel, J.: Renewable Energies in Germany's Electricity Market. Springer Netherlands, Dordrecht (2011)
16. World Energy Council: Global Energy Transitions. A Comparative Analysis of Key Countries and Implications for the International Energy Debate. Berlin (2014)
17. EEG: Gesetz für den Ausbau erneuerbarer Energien (Erneuerbare-Energien-Gesetz – EEG 2017). Erneuerbare-Energien-Gesetz vom 21. Juli 2014 (BGBl. I S. 1066), das zuletzt durch Artikel 1 des Gesetzes vom 17. Juli 2017 (BGBl. I S. 2532) geändert worden ist (2017)
18. Gartman, V., Wichmann, K., Bulling, L., Elena Huesca-Pérez, M., Köppel, J.: Wind of change or wind of challenges: implementation factors regarding wind energy development, an international perspective. AIMS Energy. (2014). https://doi.org/10.3934/energy.2014.4.485
19. Deutsche WindGuard GmbH: Status des Windenergieausbaus an Land in Deutschland 2017. im Auftrag vom Bundesverband Windenergie (BWE) und Verein Deutscher Maschinen- und Anlagenbau Power Systems (VDMA Power Systems) (2018)
20. Jessup, B.: Plural and hybrid environmental values. A discourse analysis of the wind energy conflict in Australia and the United Kingdom. Environ. Polit. (2010). https://doi.org/10.1080/09644010903396069
21. Walker, C., Baxter, J., Mason, S., Luginaah, I., Ouellette, D.: Wind energy development and perceived real estate values in Ontario, Canada. AIMS Energy. (2014). https://doi.org/10.3934/energy.2014.4.424
22. Dröes, M.I., Koster, H.R.A.: Renewable energy and negative externalities. The effect of wind turbines on house prices. J. Urban Econ. (2016). https://doi.org/10.1016/j.jue.2016.09.001
23. Reusswig, F., Braun, F., Heger, I., Ludewig, T., Eichenauer, E., Lass, W.: Against the wind. Local opposition to the German Energiewende. Util. Policy. (2016). https://doi.org/10.1016/j.jup.2016.02.006
24. Roßnagel, A., Birzle-Harder, B., Ewen, C., Götz, K., Hentschel, A., Horelt, M.-A., Huge, A., Stieß, I.: Entscheidungen über dezentrale Energieanlagen in der Zivilgesellschaft. Vorschläge zur Verbesserung der Planungs- und Genehmigungsverfahren Interdisciplinary Research on Climate Change Mitigation and Adaption, vol. 11. Kassel University Press, Kassel (2016)
25. Larsen, S.V., Hansen, A.M., Lyhne, I., Aaen, S.B., Ritter, E., Nielsen, H.: Social impact assessment in Europe. A study of social impacts in three Danish cases. J. Environ. Assess. Policy Manag. (2015). https://doi.org/10.1142/S1464333215500386
26. European Commission: Models of Horizon Scanning. How to integrate Horizon Scanning into European Research and Innovation Policies. https://www.isi.fraunhofer.de/content/dam/isi/dokumente/ccv/2015/Models-of-Horizon-Scanning.pdf (2015)
27. Amanatidou, E., Butter, M., Carabias, V., Konnola, T., Leis, M., Saritas, O., Schaper-Rinkel, P., van Rij, V.: On concepts and methods in horizon scanning. Lessons from initiating policy dialogues on emerging issues. Sci. Public Policy. (2012). https://doi.org/10.1093/scipol/scs017
28. Bulling, L., Köppel, J.: Exploring the tradeoffs between wind energy and biodiversity conservation. In: Geneletti, D. (ed.) Handbook on Biodiversity and Ecosystem Services in Impact Assessment, p. 10. Edward Elgar Publishing, Cheltenham (2016)
29. Schuster, E., Bulling, L., Köppel, J.: Consolidating the state of knowledge: a synoptical review of wind energy's wildlife effects. Environ. Manag. (2015). https://doi.org/10.1007/s00267-015-0501-5
30. Huesca-Pérez, M.E., Sheinbaum-Pardo, C., Köppel, J.: Social implications of siting wind energy in a disadvantaged region – the case of the Isthmus of Tehuantepec, Mexico. Renew. Sust. Energ. Rev. (2016). https://doi.org/10.1016/j.rser.2015.12.310
31. Gartman, V., Bulling, L., Dahmen, M., Geißler, G., Köppel, J.: Mitigation measures for wildlife in wind energy development, consolidating the state of knowledge – part 1. Planning and siting, construction. J. Environ. Assess. Policy Manag. (2016). https://doi.org/10.1142/S1464333216500137
32. Gartman, V., Bulling, L., Dahmen, M., Geißler, G., Köppel, J.: Mitigation measures for wildlife in wind energy development, consolidating the state of knowledge-part 2. Operation, decommissioning. J. Environ. Assess. Policy Manag. (2016). https://doi.org/10.1142/S1464333216500149

33. Biehl, J., Bulling, L., Gartman, V., Weber, J., Dahmen, M., Geißler, G., Köppel, J.: Vermeidungsmaßnahmen bei Planung, Bau und Betrieb von Windenergieanlagen – Synoptische Auswertungen zum Stand des Wissens. Naturschutz Landschaftsplanung. **49**(2), 63–72 (2017)
34. Akremi, L.: Stichprobenziehung in der qualitativen Sozialforschung. In: Baur, N., Blasius, J. (eds.) Handbuch Methoden der empirischen Sozialforschung, pp. 265–282. Springer VS, Wiesbaden (2014)
35. Robinson, O.C.: Sampling in interview-based qualitative research. A theoretical and practical guide. Qual. Res. Psychol. (2013). https://doi.org/10.1080/14780887.2013.801543
36. May, R., Gill, A.B., Köppel, J., Langston, R.H.W., Reichenbach, M., Scheidat, M., Smallwood, S., Voigt, C.C., Hüppop, O., Portman, M.: Future research directions to reconcile wind turbine-wildlife interactions. In: Köppel, J. (ed.) Wind Energy and Wildlife Interactions: Presentations from the CWW2015 Conference, pp. 255–276. Springer International Publishing, Cham (2017)
37. Voigt, C.C., Lehnert, L.S., Petersons, G., Adorf, F., Bach, L.: Wildlife and renewable energy. German politics cross migratory bats. Eur. J. Wildl. Res. (2015). https://doi.org/10.1007/s10344-015-0903-y
38. Grünkorn, T., Blew, J., Krüger, O., Potiek, A., Reichenbach, M., Rönn, J., von Timmermann, H., Weitekamp, S., Nehls, G.: A large-scale, multi-species assessment of avian mortality rates at land-based wind turbines in Northern Germany. In: Köppel, J. (ed.) Wind Energy and Wildlife Interactions: Presentations from the CWW2015 Conference, pp. 43–64. Springer International Publishing, Cham (2017)
39. American Wind Wildlife Institute (AWWI): Documents Library. https://awwic.nacse.org/library.php (2017). Accessed 21 Nov 2017
40. Reichenbach, M.: Wind turbines and birds in Germany – examples of current knowledge, new insights and remaining gaps. In: Köppel, J. (ed.) Wind Energy and Wildlife Interactions: Presentations from the CWW2015 Conference, pp. 239–252. Springer International Publishing, Cham (2017)
41. Länderarbeitsgemeinschaft der Vogelschutzwarten (LAG VSW): Windenergie. http://www.vogelschutzwarten.de/windenergie.htm (2017). Accessed 24 May 2017
42. Beiersdorf, A., Wollny-Goerke, K. (eds.): Ecological Research at the Offshore Windfarm "alpha ventus". Challenges, Results and Perspectives. Springer, Wiesbaden (2014)
43. Bundesamt für Seeschifffahrt und Hydrographie (BSH): Standard. Untersuchung der Auswirkungen von Offshore-Windenergieanlagen auf die Meeresumwelt (StUK4), Hamburg, Rostock (2013)
44. Bundesverwaltungsgericht (BVerwG): Beschluss vom 15. September 2009 – BVerwG 4 BN 25.09 – BRS 74 Nr. 112 (2009)
45. Bundesverwaltungsgericht (BVerwG): 9. Juli 2008, 9 A 14.07, ‚A 30, Bad Oeynhausen" (2008)
46. Fachagentur Windenergie an Land (FA Wind): Anforderungen an die planerische Steuerung der Windenergienutzung in der Regional- und Flächennutzungsplanung, Berlin. https://www.fachagentur-windenergie.de/fileadmin/files/Veranstaltungen/Dokumentation_Planerseminare_07-2016/FA_Wind_Dokumentation_Planerseminare_07-2016.pdf (2016)
47. Wolters, E.A., Steel, B.S., Lach, D., Kloepfer, D.: What is the best available science? A comparison of marine scientists, managers, and interest groups in the United States. Ocean Coast. Manag. (2016). https://doi.org/10.1016/j.ocecoaman.2016.01.011
48. Günther, M., Geißler, G., Köppel, J.: Many roads may lead to Rome: selected features of quality control within environmental assessment systems in the US, NL, CA, and UK. Environ. Impact Assess. Rev. (2017). https://doi.org/10.1016/j.eiar.2016.08.002
49. Grimm, M., Schierozek, M., Koller, M., Köppel, J., Roelcke, T.: Lesefreundliche Dokumente in Umweltprüfungen. Gefördert vom Umweltbundesamt im Rahmen des Vorhabens ‚Strategische Umweltprüfung und (neuartige) Pläne und Programme auf Bundesebene". Methoden, Verfahren, Rechtsgrundlagen, FKZ 371316100 (in press)

50. Linke, S.: Ästhetik der neuen Energielandschaften – oder: “Was Schönheit ist, das weiß ich nicht”. In: Kühne, O., Weber, F. (eds.) Bausteine der Energiewende. RaumFragen: Stadt – Region – Landschaft, pp. 409–430. Springer VS, Wiesbaden (2018)
51. Fachagentur Windenergie an Land (FA Wind): Umfrage zur Akzeptanz der Windenergie an Land – Herbst 2015, Berlin (2015)
52. Fachagentur Windenergie an Land (FA Wind): Umfrage zur Akzeptanz der Windenergie an Land – Frühjahr 2016, Berlin (2016)
53. Fachagentur Windenergie an Land (FA Wind): Umfrage zur Akzeptanz der Windenergie an Land – Herbst 2017, Berlin (2017)
54. Camargo, R., Dienes, B., Ebert, E., Günther, M., Hebrank, M., Jeong, M., Maack, A., Melilli, G., Möller-Lindenhof, T., Renner, S., Rodriguez Sotomayor, A., Rubino, M., Schniete, D., Stachnio, K., Thesen, J., Thomas, L., Tietjen, S., van Ham, F., Weber, J., Willers, A., Wüstenhagen, S., Odparlik, F.L., Köppel, J.: Social acceptance: gone with the wind? http://lehre.umweltpruefung.tu-berlin.de/mapj2016/doku.php (2016). Accessed 21 Nov 2017
55. Bundesinitiative Vernunftkraft e.V.: Vernunftkraft. http://www.vernunftkraft.de/Bundesinitiative/ (2017)
56. European Platform Against Windfarms (EPAW): European Platform Against Windfarms. http://www.epaw.org/index.php?lang=en (2008-2017). Accessed 21 Nov 2017
57. Bräuer, M., Wolling, J.: Protest oder Partizipation? Die Rolle der Bürgerinitiativen im Themenfeld Netzausbau. In: Bundesnetzagentur (BNetzA) (ed.) Wissenschaftsdialog 2015, pp. 90–103. Wirtschaft und Technologie, Kommunikation und Planung, Bonn (2015)
58. Devine-Wright, P.: Rethinking NIMBYism. The role of place attachment and place identity in explaining place-protective action. J. Community Appl. Soc. Psychol. (2009). https://doi.org/10.1002/casp.1004
59. Wolsink, M.: Wind power and the NIMBY-myth. Institutional capacity and the limited significance of public support. Renew. Energy. (2000). https://doi.org/10.1016/S0960-1481(99)00130-5
60. Petrova, M.A.: From NIMBY to acceptance: toward a novel framework – VESPA – for organizing and interpreting community concerns. Renew. Energy. **86**, 1280–1294 (2016)
61. van der Horst, D.: NIMBY or not? Exploring the relevance of location and the politics of voiced opinions in renewable energy siting controversies. Exploring the relevance of location and the politics of voiced opinions in renewable energy siting controversies. Energy Policy. (2007). https://doi.org/10.1016/j.enpol.2006.12.012
62. Brunnengräber, A.: Klimaskeptiker im Aufwind. Wie aus einem Rand ein breiteres Gesellschaftsphänomen wird. In: Kühne, O., Weber, F. (eds.) Bausteine der Energiewende. RaumFragen: Stadt – Region – Landschaft, pp. 271–292. Springer VS, Wiesbaden (2018)
63. Bundesinitiative Vernunftkraft e.V.: Positionen für mehr Weitsicht zum Wohl von Mensch und Natur. http://www.vernunftkraft.de/de/wp-content/uploads/2014/07/Positionspapier-Juli-2014.pdf (2014)
64. Eichenauer, E., Reusswig, F., Meyer-Ohlendorf, L., Lass, W.: Bürgerinitiativen gegen Windkraftanlagen und der Aufschwung rechtspopulistischer Bewegungen. In: Kühne, O., Weber, F. (eds.) Bausteine der Energiewende. RaumFragen: Stadt – Region – Landschaft, pp. 633–652. Springer VS, Wiesbaden (2018)
65. Roßmeier, A., Weber, F., Kühne, O.: Wandel und gesellschaftliche Resonanz – Diskurse um Landschaft und Partizipation beim Windkraftausbau. In: Kühne, O., Weber, F. (eds.) Bausteine der Energiewende. RaumFragen: Stadt – Region – Landschaft, pp. 653–680. Springer VS, Wiesbaden (2018)
66. Sullivan, P.J., Acheson, J.M., Angermeier, P.L., Faast, T., Flemma, J., Jones, C.M., Knudsen, E.E., Minello, T.J., Secor, D.H., Wunderlich, R., Zanatell, B.A.: Defining and Implementing Best Available Science for Fisheries and Environmental Science, Policy, and Management. http://fisheries.org/docs/policy_science.pdf (2006). Accessed 27 June 2017
67. Ryan, C.M., Cerveny, L.K., Robinson, T.L., Blahna, D.J.: Implementing the 2012 forest planning rule: best available scientific information in forest planning assessments. For. Sci. (2018). https://doi.org/10.1093/forsci/fxx004

68. Länderarbeitsgemeinschaft der Vogelschutzwarten (LAG VSW): Abstandsempfehlungen für Windenergieanlagen zu bedeutsamen Vogellebensräumen sowie Brutplätzen ausgewählter Vogelarten (Stand April 2015). http://www.vogelschutzwarten.de/downloads/lagvsw2015_abstand.pdf (2015). Accessed 23 May 2017
69. Bundesverwaltungsgericht (BVerwG): BVerwG Urteil vom 21.11.2013, 7 C 40.11 (2013)
70. Williams, B.K., Brown, E.D.: Adaptive management: from more talk to real action. Environ. Manag. (2014). https://doi.org/10.1007/s00267-013-0205-7
71. Hanna, L., Copping, A., Geerlofs, S., Feinberg, L., Brown-Saracino, J., Bennet, F., May, R., Köppel, J., Bulling, L., Gartman, V.: Results of IEA Wind Adaptive Management White Paper. International Energy Agency Wind Implementing Agreement (2016)
72. Bulling, L., Köppel, J.: Adaptive Management in der Windenergieplanung – Eine Chance für den Artenschutz in Deutschland? Naturschutz Landschaftsplanung. **49**(2), 73–79 (2017)
73. Wulfert, K., Lau, M., Widdig, T., Müller-Pfannenstiel, K., Mengel, A.: Standardisierungspotenzial im Bereich der arten- und gebietsschutzrechtlichen Prüfung. FuE – Vorhaben im Rahmen des Umweltforschungsplanes des Bundesministeriums für Umwelt, Naturschutz und Reaktorsicherheit im Auftrag des Bundesamtes für Naturschutz – FKZ 3512 82 2100 (2015). Accessed 16 May 2017
74. Brandt, E.: Gutachterliche Stellungnahme zu ausgewählten Fragestellungen im Zusammenhang mit der geplanten Novellierung des Bundesnaturschutzgesetzes. im Auftrag des Fördervereins der Koordinierungsstelle Windenergierecht (k:wer) e. V. (2017)
75. Peste, F., Paula, A., da Silva, L.P., Bernardino, J., Pereira, P., Mascarenhas, M., Costa, H., Vieira, J., Bastos, C., Fonseca, C., Pereira, M.J.R.: How to mitigate impacts of wind farms on bats? A review of potential conservation measures in the European context. Environ. Impact Assess. Rev. (2015). https://doi.org/10.1016/j.eiar.2014.11.001
76. Sims, C., Hull, C., Stark, E., Barbour, R.: Key learnings from ten years of monitoring and management interventions at the bluff point and Studland Bay wind farms: results of a review. In: Hull, C., Bennett, E., Stark, E., Smales, I., Lau, J., Venosta, M. (eds.) Wind and Wildlife: Proceedings from the Conference on Wind Energy and Wildlife Impacts, October 2012, Melbourne, Australia, pp. 125–144. Springer Netherlands, Dordrecht (2015)
77. Grünkorn, T., Rönn, J. von Blew, J., Nehls, G., Weitekamp, S., Timmermann, H., Reichenbach, M., Coppack, T., Potiek, A., Krüger, O.: Ermittlung der Kollisionsraten von (Greif-)Vögeln und Schaffung planungsbezogener Grundlagen für die Prognose und Bewertung des Kollisionsrisikos durch Windenergieanlagen (PROGRESS). Schlussbericht zum durch das Bundesministerium für Wirtschaft und Energie (BMWi) im Rahmen des 6. Energieforschungsprogrammes der Bundesregierung geförderten Verbundvorhaben PROGRESS, FKZ 0325300A-D (2016). Accessed 19 Aug 2016
78. U.S. Fish and Wildlife Service Midwest Region der Staaten von Iowa, Illinois, Indiana, Michigan, Minnesota, Missouri, and Wisconsin and the American Wind Energy Association: Midwest Wind Energy. Multi-Species Habitat Conservation Plan. Public Review Draft (2016)
79. National Wind Coordinating Collaborative (NWCC): Greater Sage-Grouse. Overview and Effects of Wind Energy Development (2017)
80. Sudfeldt, C., Dröschmeister, R., Frederking, W., Gedeon, K., Gerlach, B., Grüneberg, C., Karthäuser, J., Langgemach, T., Schuster, B., Trautmann, S., Wahl, J.: Vögel in Deutschland – 2013, Münster (2013)
81. Department of the Interior (DOI), US Fish and Wildlife Service (USFWS): Eagle Permits; Revisions to Regulations for Eagle Incidental Take and Take of Eagle Nests. 50 CFR Parts 13 and 22, vol. 81 (2016)
82. U.S. Fish & Wildlife Service: Bald and Golden Eagles: Population demographics and estimation of sustainable take in the United States, 2016 update, Washington, DC (2016)
83. Diffendorfer, J., Beston, J., Merrill, M., Stanton, J., Corum, M., Loss, S., Thogmartin, W., Johnson, D., Erickson, R., Heist, K.: A method to assess the population-level consequences of wind energy facilities on bird and bat species. In: Köppel, J. (ed.) Wind Energy and Wildlife Interactions: Presentations from the CWW2015 Conference, pp. 65–76. Springer, Cham (2017)

84. Moore, J.E., Curtis, K.A., Lewinson, R.L., Dillingham, P.W., Cope, J.M., Fordham, S.V., Heppell, S.S., Pardo, S.A., Simpfendorfer, C.A., Tuck, G.N., Zhou, S.: Evaluating sustainability of fisheries bycatch mortality for marine megafauna. A review of conservation reference points for data-limited populations. Environ. Conserv. (2013). https://doi.org/10.1017/S037689291300012X
85. O'Brien, S.H., Cook, A.S.C.P., Robinson, R.A.: Implicit assumptions underlying simple harvest models of marine bird populations can mislead environmental management decisions. J. Environ. Manag. (2017). https://doi.org/10.1016/j.jenvman.2017.06.037
86. Pahl-Weber, E., Henckel, D.: The Planning System and Planning Terms in Germany. A Glossary, p. 7. Studies in Spatial Development, Hannover (2008)
87. Raschke, M.: Die Windenergieerlasse der Länder. In: Müller, T., Kahl, H. (eds.) Energiewende im Föderalismus. Schriften zum Umweltenergierecht, vol. 18, pp. 261–289. NOMOS, Baden-Baden (2015)
88. Christ, B., Linke, H.J.: Windenergieanlagen in der Raumordnung – landes- und regionalplanerische Ansätze. Flächenmanagement und Bodenordnung (fub)(1) (2014)
89. Nagel, P.-B., Schwarz, T., Köppel, J.: Ausbau der Windenergie – Anforderungen aus der Rechtsprechung und fachliche Vorgaben für die planerische Steuerung. Umweltmed. Planungsrecht. **10**, 371–382 (2014)
90. Gailing, L.: Die Landschaften der Energiewende – Themen und Konsequenzen für die sozialwissenschaftliche Landschaftsforschung. In: Gailing, L., Leibenath, M. (eds.) Neue Energielandschaften – Neue Perspektiven der Landschaftsforschung. RaumFragen: Stadt – Region – Landschaft, pp. 207–215. Springer VS, Wiesbaden (2013)
91. Noori, M., Kucukvar, M., Tatari, O.: Economic input–output based sustainability analysis of onshore and offshore wind energy systems. Int. J. Green Energy. (2015). https://doi.org/10.1080/15435075.2014.890103
92. Huang, Y.-F., Gan, X.-J., Chiueh, P.-T.: Life cycle assessment and net energy analysis of offshore wind power systems. Renew. Energy. (2017). https://doi.org/10.1016/j.renene.2016.10.050
93. Hooper, T., Beaumont, N., Hattam, C.: The implications of energy systems for ecosystem services. A detailed case study of offshore wind. Renew. Sust. Energ. Rev. (2017). https://doi.org/10.1016/j.rser.2016.11.248
94. Kraemer, A., Carin, B., Gruenig, M., Naves Blumenschein, F., Flores, R., Mathur, A., Brandi, C., Spencer, T., Helgenberger, S., Thielges, S., Vaughan, S., Whitley, S., Ruet, J., Ott, H.: Green Shift to Sustainability: Co-Benefits & Impacts of Energy Transformation on Resource Industries, Trade, Growth, and Taxes. G20 Insights (2017)
95. Kraemer, A.: Green shift to sustainability: co-benefits and impacts of energy transformation. Policy Brief, 109 (2017)
96. May, R.: Mitigation for birds. In: Perrow, M. (ed.) Wildlife and Wind Farms – Conflicts and Solutions. Volume 1 Onshore: Potential Effects, pp. 124–144. Pelagic Publishing, Exeter (2017)
97. Bulling, L., Sudhaus, D., Schnittker, D., Schuster, E., Biehl, J., Tucci, F.: Vermeidungsmaßnahmen bei der Planung und Genehmigung von Windenergieanlagen. Bundesweiter Katalog von Maßnahmen zur Vermeidung des Eintritts von artenschutzrechtlichen Verbotstatbeständen nach §44 BNatSchG (2015). Accessed 19 Aug 2016
98. Fachagentur Windenergie an Land (FA Wind): Kompensation von Eingriffen in das Landschaftsbild durch Windenergieanlagen im Genehmigungsverfahren und in der Bauleitplanung. Berlin (2016)
99. Länderausschuss für Immissionsschutz (LAI): Hinweise zur Ermittlung und Beurteilung der optischen Immissionen von Windenergieanlagen (WEA-Schattenwurf-Hinweise). verabschiedet auf der 103. Sitzung (2002)
100. European Commission, European Parliament: Directive 2000/60/EC of the European Parliament and of the Council of 23 October 2000 establishing a framework for Community action in the field of water policy. Water Framework Directive (2000)

101. May, R., Hamre, Ø., Vang, R., Nygård, T.: Evaluation of the DTBird video-system at the Smøla wind-power plant. Detection capabilities for capturing near-turbine avian behaviour. NINA Report 910 (2012)
102. Hanagasioglu, M., Aschwanden, J., Bontadina, F., de la Puente Nilsson Marcos: Investigation of the effectiveness of bat and bird detection of the DTBat and DTBird systems at Calandawind turbine. Final report (2015)
103. Litsgård, F., Eriksson, A., Wizelius, T., Säfström, T.: DTBird system Pilot Installation in Sweden. Possibilities for bird monitoring systems around wind farms. Experiences from Sweden's first DTBird installation. Ecocom AB. 21-12-2016. Englische Übersetzung durch DTBird (2017). Accessed 7 Mar 2017
104. Department of Energy (DoE): 3 Ways Energy Department Research Will Help Coexist with Wind Energy Deployment. https://energy.gov/eere/articles/3-ways-energy-department-research-will-help-eagles-coexist-wind-energy-deployment (2016)
105. DTBird: U.S. Department of Energy Funds Evaluation of DTBird System. http://www.dtbird.com/index.php/es/news-2/item/43-usa-department-energy-evaluate-effectiveness-dtbird-system (2016)
106. Miller, M., Giebel, R.: Rotor-mounted bat impact deterrence system (poster). In: Wind Wildlife Research Meeting XI: Presentation Abstracts, pp. 52–53 (2016)
107. Bat Conservation International: Ultrasonic Acoustic Deterrents (UADs). http://www.batcon.org/our-work/regions/usa-canada/wind2/ultrasonic (2017)
108. Bayerischer Verwaltungsgerichtshof: Urteil vom 10.03.2016, Az. 22 B 14.1875 und 22 B 14.1876 (2016). Accessed 23 Feb 2017
109. Bundesamt für Naturschutz (BfN): NATforWINSENT – Naturschutz im Windtestfeld. Entwicklung eines Konzepts zur Naturschutzbegleitforschung im Rahmen des WindForS-Windenergietestfelds Schwäbische Alb. FKZ 3517 86 1600. https://www.natur-und-erneuerbare.de/projektdatenbank/projekte/natforwinsent-naturschutz-im-windtestfeld/ (2017)
110. Scottish Natural Heritage (SNH): Guidance on Methods for Monitoring Bird Populations at Onshore Wind Farms. http://www.snh.gov.uk/docs/C205417.pdf (2009). Accessed 11 July 2017
111. Ornitela: Ornithology and Telemetry Applications. http://www.ornitela.com/ (2016)
112. Moberg, F., Simonsen, S.H.: What is resilience? An introduction to social-ecological research (2014)
113. Rockström, J., Falkenmark, M., Folke, C., Lannerstad, M., Barron, J., Enfors, E., Gordon, L., Heinke, J., Hoff, H., Pahl-Wostl, C.: Water Resilience for Human Prosperity. Cambridge University Press, Cambridge (2014)
114. Reed, M., Evely, A., Cundill, G., Fazey, I., Glass, J., Laing, A., Newig, J., Parrish, B., Prell, C., Raymond, C., Stringer, L.: What is social learning? Ecol. Soc. (2010). https://doi.org/10.5751/ES-03564-1504r01
115. Bodin, Ö.: Collaborative environmental governance. Achieving collective action in social-ecological systems. Science (New York, N.Y.). (2017). https://doi.org/10.1126/science.aan1114
116. Scott, T., Thomas, C.: Unpacking the collaborative toolbox: why and when do public managers choose collaborative governance strategies? Policy Stud. J. **45**(1), 191–214 (2017)
117. Wondolleck, J., Yaffee, S.: Making Collaboration Work. Lessons from Innovation in Natural Resource Management. Island Press, Washington, DC (2000)
118. Davies, W.: The age of post-truth politics. The New York Times (26 August 2016). https://www.nytimes.com/2016/08/24/opinion/campaign-stops/the-age-of-post-truth-politics.html
119. Krane, M.: Alternative Fakten kann jeder. Fake News entlarven. Der Tagesspiegel (20 June 2017). http://www.tagesspiegel.de/themen/freie-universitaet-berlin/fake-news-entlarven-alternative-fakten-kann-jeder/19938682.html

The Mitigation of Impact and the Impact of Mitigation: An Ethical Perspective

Roel May

With birds in its wake
Painted wings, feathered blades
Art of reduction

Abstract Societal concerns regarding the negative impacts of wind turbines on species and ecosystems have placed more emphasis on mitigation efforts pre- and post-construction. While the mitigation hierarchy is usually fronted to deal with negative ecological impacts, it is hardly employed accordingly. This calls for the core of the problem to be addressed, namely, the lack of an appropriate framework for mitigation as a concept to properly address ecological impacts caused by wind-power development. In this chapter, mitigation is defined as the intervention(s) implemented to affect the risk of wind turbines impacting species or ecosystems. This concept is placed within a social-ecological context where the consecutive steps of the mitigation hierarchy may be affected by socio-economic, technological or environmental spheres of interest. Decisions relating to mitigation are in principle normative, which necessitates addressing three central ethical questions: (1) In which circumstances should mitigation be implemented? (2) How much mitigation is required? (3) Who is responsible for mitigation? Implementing mitigation requires decision-makers to acknowledge that trouble never comes alone, which requires balancing trade-offs and embracing uncertainty into the decision-making process. Adaptive and participatory management may be the best decision-making framework to do this, as it allows for improved ecological understanding through monitoring and a flexible approach to mitigate locally but manage regionally.

Keywords Wind-power development · Wildlife impacts · Social-ecological systems · Intervention ecology · Risk · Impact significance · Precautionary

---

R. May (✉)
Norwegian Institute for Nature Research (NINA), Trondheim, Norway
e-mail: Roel.May@nina.no

R. Bispo et al. (eds.), *Wind Energy and Wildlife Impacts*,
https://doi.org/10.1007/978-3-030-05520-2_6

principle · Decision-making framework · Environmental ethics · Mitigation hierarchy

1 Introduction

With the increasing development of wind power, it has become increasingly difficult to locate wind-power plants in areas without ecological conflicts. Societal concerns for the negative impacts of wind turbines on species and ecosystems have placed more emphasis on mitigation efforts pre- and post-construction. While the mitigation hierarchy is usually fronted to deal with negative ecological impacts, it is hardly employed accordingly [1–4]. With the maturing of the wind-power industry, new research and development has also led to technological innovations to mitigate impacts. While technology can contribute to mitigating impacts, it should however not be the default rationale. The belief that human ingenuity through the reliance on technological measures can 'fix' ecological impacts caused by anthropogenic activities has been criticized [5]. Reliance on techno-fixes enhances the risk of only dealing with those impacts that can be technically solved, while less obvious impacts remain under the radar [6]. Another perspective is that economic cost-benefit analyses form the backdrop of many decisions related to mitigation effort. The marginal financial benefits of wind-power projects and unclear permitting requirements are used as reasons for not setting requirements for mitigation during the application process or refraining from mitigation post-consent [2, 7]. This could be questioned because the negative ecological effects a project may have were unlikely part of the budget in the first place. In that sense ecological impacts are public costs external to the private costs of developing wind power [8]. However, also the uncertain costs related to the risk of an ecological impact versus the equally uncertain potential benefits through a reduction in CO_2 emissions complicate decision-making. It would however be too easy just to blame engineers and economists [9]. It rather calls for addressing the core of the problem, being the lack of an appropriate framework for mitigation as a concept to properly address ecological impacts caused by anthropogenic activities such as wind-power development and to provide additional opportunities for nature [7, 10]. Conceptualizing mitigation will help overcome procedural and normative conflicts and enhance responsibility for sustainable development [9, 11].

2 Mitigation Concept

2.1 Defining Mitigation

If we wish to understand what mitigation conceptually means, it is important to know its definition. As there exists no generally accepted definition for mitigation,

I will use the following definition: 'Mitigation relates to the intervention(s) implemented to affect the risk of wind turbines impacting species or ecosystems'. This definition includes three central components that require further elaboration: impact, risk and intervention.

'Impact' can have various connotations and is often confused and intermixed with the word 'effect'. It is however important to clarify the difference between these two terms. While an effect pertains to the number of individuals of a species or quantity of ecosystem components being affected by wind turbines, an impact indicates a *significant* effect on a species' survival or on ecosystem integrity, respectively [12]. Effects of wind turbines on wildlife are generally due to collisions, disturbance, barriers and habitat loss. The nature, magnitude and spatial extent of these effects are however highly species-, site- and condition-specific [13]. The extent to which a species is affected is dependent on its sensory perception abilities, aero- or aquadynamics, behaviour and life history. For instance, a certain number of birds or bats colliding with wind turbines may or may not lead to an impact on their respective populations or impact ecosystem functioning. Still, in one specific case effect and impact may actually be synonyms. Taking an individual approach, any individual that is affected will automatically also be impacted. An example of the individual approach would be 'incidental take permits' for the golden eagle (*Aquila chrysaetos*) [14], which is protected under both the Bald and Golden Eagle Protection Act and the Migratory Bird Treaty Act in the United States of America. Similarly, under the EU legislation (Habitats and Birds Directives) as well as international conventions (EUROBATS, Bern and Bonn Conventions), deliberate killing, disturbance or deterioration of sites for species of 'community interest' (including (migratory) microbats and raptors) is in principle prohibited, unless for imperative reasons of overriding public interest and in selective strictly supervised cases [cf. 15]. Most commonly however, a population approach is taken whereby impacts depend on a species' population status and life history [15, 16]. Given a biogeographical distribution that usually exceeds the footprint of a wind-power plant by an order of magnitude, impacts on populations may be hard to determine at the project scale. Whether an impact will occur at the ecosystem level depends on the trophic position of a species in the food web [17, 18]. In general, higher levels of organization make it more difficult to assess impacts caused by a specific wind-power project properly [12, 19].

The second component of the mitigation definition relates to risk. Risk is usually defined as the probability of harm (i.e. hazard) to occur due to exposure and vulnerability to harmful events [20]. Vulnerability relates to a species' sensitivity to change (i.e. lack of resistance) and its resilience when change occurs due to its capacity to cope (i.e. tolerance) and/or to adapt to wind turbine-induced effects causing an impact [cf. 20]. In other words, it pertains to both the sensitivity of a species to wind turbines and its consequence once it occurs. This specifically relates to the effects disturbance or collisions can have on a species' demography [16]. What is important to understand is the difference between risk and uncertainty. While risk refers to known metrics of probability and exposure to harmful event (e.g. through dose-response relationships), uncertainty relates to incomplete knowledge or even ignorance about the outcomes of such an event [21]. While risk may have statistical uncertainty, the effects are known and probabilities quantifiable [22]. However, due to lack of knowledge, risks may become unquantifiable whereby events have uncertain cause-effect responses and therefore unknown probabilities. A third form of uncertainty may arise when there exists disagreement on the extent of an impact, for instance, relating to green versus green dilemmas [3]. Proper impact assessments therefore need to be based on ecologically meaningful and competing models with good predictive power [21].

The third component is 'intervention', which can be defined as any action to interfere with the outcome of a condition or process to prevent harm or improve functioning. Interventions can consist of proactive (avoid, minimize), active (reduce) and reactive (compensate, restore) actions at any spatial scale [23]. An intervention not only reflects specific measures but also the act of intervening. This is the realm of the research discipline of intervention ecology, which aims at reconciling conservation biology and restoration ecology [23, 24]. Typically, conservation biology aims at proactively preserving natural areas from impacts (precaution). However, 'true' wilderness is mostly gone due to widespread anthropogenic activities over long timeframes. Restoration ecology instead aims at reactively re-creating historic natural conditions in degraded areas (post-caution). It is however utopic to assume it can re-create these due to shifting baselines [25]. Intervention ecology intends to maintain the current desired ecological state or improve a current undesirable state [23]. It aims at targeting so-called leverage points: key components or flows that either stabilize or provoke rapid change in ecosystems. Interventions may encompass all from technical, ecological and socio-economic actions, as well as deliberate non-interventions (i.e. decisions not to act). Hobbs and Hallet [23] suggest that changing rules and governance at multiple scales may be more effective than 'tinkering with ecosystem properties per se'. Intervening thus requires good insight into the social-ecological system and knowledge on the aspects that give most leverage for mitigation. While conservation biology will invariantly put more weight on avoiding or minimizing impacts, restoration ecology will be directed at compensation and restoration. Intervention ecology can reconcile these two disciplines and address the entire mitigation hierarchy.

2.2 Mitigation Framework

Mitigation can be implemented at different phases during the development of wind-power plants, from strategic planning through to decommissioning [26]. During any of these phases, specific decision-making processes are at play, covering economic, technological, social, ecological and ethical spheres of interest. For mitigation to be considered, both socio-economic interest for technological development of wind power and societal concern for ecological impacts due to the development are required. Within this perspective, mitigation can be placed within the concept of social-ecological systems [27, 28], which recognizes the interlinkages existing between humans and their environment. This may include societal drivers such as governance, technological resource use such as wind turbines and natural resources

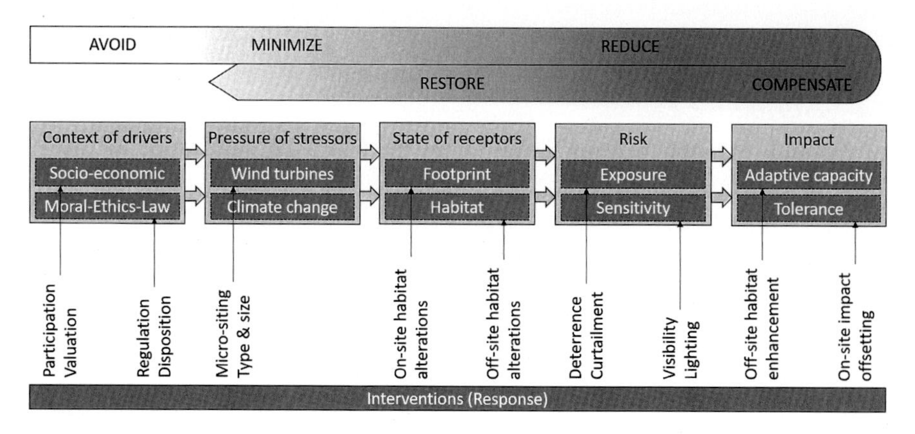

Fig. 1 Conceptual framework decomposing central mitigation components and possible measures following to the DPSIR risk framework

(e.g. birds or bats). To better visualize the interactions between society and the environment, the European Environment Agency promoted the Drivers-Pressures-States-Impacts-Responses (DPSIR) framework [29]. To also include precautionary perspectives, this framework can be extended with the risk assessment framework [30]. The combined framework then links the Context derived from driving forces (e.g. reducing carbon emissions through development of wind power) to the Pressure caused by stressors (e.g. wind turbines, access roads) on the State of receptors (e.g. raptors, bats, marine mammals), leading to the Risk of an impact (e.g. collisions, disturbance). When conceptualizing the mitigation process within these associated frameworks (Fig. 1), it becomes clear that different components play a role during the various phases of the mitigation hierarchy. Socio-economic and regulatory context and planned wind-power plants as consequent stressors are central drivers during the avoidance phase. Minimization aims at mitigating the potential pressure of the wind turbines prior to construction, for instance, through micrositing offset against the potential loss of producing emission-free energy. Reduction measures target the risk of an impact and are mostly affected by the expected efficacy of the proposed mitigation measures. A species' need for conservation action, but also societal concern, guides the extent of compensation during operation. Finally, restoration aims at improving the state of the affected area and/or species upon decommissioning. This framework can thus reveal how the different mitigation phases require different decision-making approaches and ethics. A variety of interventions target the different components of the framework and thereby the social-ecological system as such (Fig. 1).

3 Mitigation Ethics

3.1 Normative Ethics

Decisions relating to the mitigation of environmental impacts caused by wind turbines have economic, technological, social and environmental dimensions that need to be satisfied. In order to make the right decisions, decision-makers have to reconcile potentially conflicting dimensions by identifying actions that can help overcome these conflicts [12]. What defines the *right* decision is the field of normative ethics. In addition, mitigation decisions relate to human interactions with the natural environment. Environmental ethics is the study of normative issues and principles that guide our actions and policies across the entire range of environmental issues [31]. Normative and environmental ethics can therefore help understand the principles of mitigation actions. Normative ethics reflects on and provides criteria for how one should, or ought to, act morally speaking. Such criteria can be based on utilitarian (or consequentialist), deontological or virtue theory. Utilitarianism focuses on the expected outcome of actions (i.e. its value, goodness, desirability) based for instance on predictive models [32] or cost-benefit analyses [33, 34]. Deontology focuses on the actions themselves evaluated against a set of principle rules and may include guidelines [35] or multi-criteria decision analyses [36, 37]. Virtue ethics focuses on the attitudes (i.e. virtues and vices such as care, temperance or greed) leading to the actions [38, 39], such as good practice guidance [40]. Their points of departure therefore differ with respect to an action.

Throughout a wind-power projects' life cycle, more certainty can be obtained on the (potential) impact caused by turbines on wildlife. Decisions pertaining to mitigation actions are therefore based on different ethics throughout the mitigation hierarchy. Avoidance as a proactive measure during the planning phase, when uncertainty on the potential impact is high, is generally based on deontological ethics with a focus on the development itself [38]. Rules set out in regulations and guidelines [e.g. 41, 42] define the boundaries within which development is deemed allowable. Often multi-criteria decision assessment approaches are used to assess possible siting alternatives to avoid negative impacts on the environment and local communities while ensuring development of renewable energy [26, 43]. All other types of mitigation are generally based on utilitarian ethics which focus on the expected outcomes of an action [38]. Examples of this are repowering to enable resiting different types of turbines to obtain less impact per kWh [44] or curtailment actions that balance temporal shutdown to reduce an impact with the loss of production [45]. While minimization and reduction represent active measures during the design and construction phase, respectively, compensation and restoration are reactive measures during the operation and decommissioning phase. Compensating known impacts may have ethical implications that differ from the other types of mitigation which are strictly aimed at the impacts at the wind-power plant site. Compensation can both include on-site and off-site, as well as in-kind and out-of-kind measures [26]. Especially (monetary) offsetting (i.e. off-site out-of-kind compensation) sees nature as a commodity that can be substituted [38]. This may be possible regarding the instrumental values of biodiversity and ecosystems, i.e. ecosystem services, as these represent the commodities of nature for humans [cf. 10]. However, endangered species that go locally extinct cannot be replaced per se by other species elsewhere. As time progresses, impacts may accrue causing most harm in the future. A pure utilitarian ethic would however underestimate future impacts due to discounting [46–48], eroding our obligation to protect biodiversity [38]. Due to such considerations, compensation is usually seen as a last resort to mitigate residual impacts. Restoration at decommissioning is generally based on the utilitarian assumption that the impacted wind-power plant footprint can be restored to preconstruction conditions. However, restoration ecology also includes a virtue ethical component as it aims to engage people into the practice of assisted recovery of an ecosystem [25, 35].

So, to make the *right* decisions regarding the implementation of mitigation actions requires insight into the ethical principles guiding decision-making. The concept of energy justice has recently emerged to apply social justice principles to energy systems [9, 49]. The energy justice framework consolidates three core considerations: distributional justice, procedural justice and recognition justice [49]. For a decision-maker, three ethical questions will be central with regard to the implementation of mitigation measures: (1) In which circumstances should mitigation be implemented? (2) How much mitigation is required? (3) Who is responsible for mitigation? In the following section, I will try to reflect on each of these three questions.

3.2 In Which Circumstances Should Mitigation Be Implemented?

Obviously, the answer to this question should read: when there is a significant (potential) impact. Duinker [50] identified four categories of impact significance: statistical, ecological, social and decision-related significance. Statistical significance relates to isolation of project-induced changes from natural variation, whereas ecological significance relates to the importance of project-induced changes from an ecological perspective. Social and decision-related significance, however, are derivates of the actual project-induced impact. Social significance here relates to the acceptability of project-induced changes to the environment and decision-related significance to whether project-induced changes influence financial or technical decision-making [50]. Given these four categories, impact significance clearly is normative in character and context dependent [51]. When its rationale is not clearly defined, an impact may either encompass species with a clear social significance (e.g. white-tailed eagle safaris, hunting ptarmigan), keystone species with an important role in the ecosystem, or statistically affected species whose impact can be mitigated [e.g. bat curtailment: 52, 53]. Clearly defining which impacts are to be considered, and how, should therefore be the starting point of significance determinations [51].

If there exists uncertainty about the significance of an impact, my question may be more difficult to answer. In cases of incomplete knowledge on a potential impact caused by a wind-power development, risk-based approaches may not be warranted as those assume that both the exposure and probability of harm can be quantified. Although statistical uncertainty of the actual predictions may be allowed, the basic cause-effect responses are thought to be known. Often, however, our understanding of how exactly wind turbines affect a population through changes in demographic characteristics is limited [54]. Population impacts are highly species- and site-specific and vary in time and space [16]. Decision-making under uncertainty has therefore focused on more qualitative approaches to deal with uncertain impacts, such as the precautionary principle. The principle's rationale is that scientific uncertainty is not a reason to refrain from preventive action against adverse environmental (risk of) impacts [22, 55]. The precautionary principle has been criticized much, not in the least for being overly precautionary, hampering development [55, 56]. Still it is implemented in many countries' legislation. I will not go into the actual debate surrounding the precautionary principle but merely point out three important criteria of the principle. The precautionary principle firstly combines scientific uncertainty with the plausibility of the impact to occur. This inherently sets minimum thresholds regarding the likelihood ('reasonable ground for concern') and seriousness ('severity and/or irreversibility') of anticipated harm [55]. Still, depending on your perspective, there may still be dispute whether or not (the risk of) an impact is likely and serious enough. Therefore, the proportionality criterion aims to correspond the plausibility to required precautionary action. Hermerén [57] identified four conditions for proportionality that need to be addressed, which – when applied to mitigation – are *importance* of mitigating an impact, *relevance* of the proposed mitigation measure to bring this about, *most favourable option* among alternative measures (e.g. least uncertainty or habituation) and *non-excessiveness* of the proposed mitigation measure relative to the impact (e.g. cost-effectiveness of the measure or whether implementation causes unwarranted delays in development relative to the intensity and type of impact). Dependent on the consistency of the impact and the efficiency of the mitigation action, this may have various outcomes [21]. When having to decide relative to the expected range of risk of an impact to occur (ranging from best to worst possible outcome), either the likelihood or the exposure to this risk can be mitigated. Whether to mitigate relative to the best or worst possible outcome (e.g. the number of white-tailed eagle collisions at the Smøla wind-power plant ranges between 2 and 11 per year) may depend on the vulnerability of the species involved. While for threatened species mitigation actions should be implemented to reduce the likelihood of any impact to occur (mitigating the best outcome; reduce the minimum of two collisions per year), for species of less concern, the exposure can be mitigated (mitigating the worst outcome; reduce the maximum of 11 collisions per year). These two perspectives provide extremes against which to decide which alternative to implement. These two criteria have been reconciled into the 'reality criterion' which, based on a decision-maker's optimism, may tend more towards the one or the other before-mentioned extremes [58]. This criterion may well be relevant for decision-makers to balance impact consistency against efficiency of mitigation [56].

3.3 How Much Mitigation Is Required?

When mitigation is required, the extent to which this needs to be implemented depends on the efficacy of specific measures to mitigate impact significance [1, 12, 59–61]. Impacts should thereby be mitigated according to the mitigation hierarchy until they become insignificant [26]. The no-net-loss principle is promoted to ensure that development does not lead to net loss, but rather net gain, of biodiversity [8, 62]. However, it remains unclear what this really entails. The International Financing Corporation defined no-net-loss as: "The point at which the project-related impacts on biodiversity are balanced by measures taken to avoid and minimize the project's impacts, to understand on site restoration and finally to offset significant residual impacts, if any, on an appropriate geographic scale (e.g. local, landscape-level, national, regional)" [63]. Adopting this definition, clearly only *impacts* need to be offset, following a population approach, for instance, pertaining to the point at which a species' demography is no longer affected by the development. How a *gain* can be obtained in this sense remains unclear, as that would imply that first all *effects* need to be nullified before a population can gain a positive trajectory. Another unclarity is related to what *net* is referring to. One perspective would be that no-net-loss occurs when the post-construction impacts at the wind-power plant site are mitigated such that a species can live as if in a preconstruction situation. However, what is fair to include? Should beside direct impacts (wind turbines, roads, power lines) also indirect impacts external to the project (e.g. increased public access to the general area) be included? Does *net* also include positive effects on ecosystems by reducing CO_2 emissions or providing novel habitats such as artificial reefs at subsea turbine foundations [2, 64, 65]? If so, it may become difficult to assess what the net impacts are and how much these should be mitigated to reach no-net-loss [but see, e.g. 66]. Given the inherent uncertainty in the (potential) impact as well as in the (expected) efficacy of mitigation measures, it may be difficult to assess the extent of any residual impact after having implemented a measure [2].

Irrespective of the no-net-loss principle, the question remains how much of an impact may remain after mitigation. Although utilitarian reasoning can incorporate mitigation to enhance total benefit and recommend only those alternatives comparatively least harmful, this ethical framework can similarly justify sacrifices for the greater welfare if mitigation and alternatives are infeasible or extremely costly [67]. Policy favouring public land sites for renewable energy facilities may, for instance, justify incurring some regional impediments for the sake of national interests, as in Vermont [67].

3.4 Who Is Responsible for Mitigation?

Again, this question may seem easy to answer: the developer or operator should as 'polluter' pay for mitigation of wildlife impacts. However, as wildlife impacts are not intended, these become public costs external to the private costs of developing wind power [8]. The polluter pays principle aims at internalizing such external costs of 'pollution' to developers to prevent harm to the environment [8, 68]. From a utilitarian (e.g. economic) viewpoint, the danger exists that complete result orientation releases corporations and governments from personal responsibilities [67]. This may be strengthened by compartmentalization (e.g. permitting authorities may not be responsible for the environment) causing moral distancing; while feeling responsible for the results, they tend to overlook their accountability for other foreseeable results of their actions (unintended by-products) [67]. Such effects may be countered by corporate governance and business strategies [69, 70], such as adhering to international standards (e.g. IFC Performance Standard 6) [41, 63].

However, due to the intertwined roles stakeholders have, it will be more difficult to establish who the polluter is [71, 72]. Stabell and Steel [68] developed an ethical framework for the fair distribution of costs for precautionary measures, based on the polluter pays and ability to pay principles. Their framework is intended to help decision-makers think more systematically about distributional consequences of taking precautionary measures. Ringius and Torvanger [73] assessed fairness of burden sharing in international climate policy based on four principles of equity: responsibility (for development), need (for wildlife), contribution (to mitigation) and capacity (ability to pay). Because wind power is developed to help mitigate climate change, and because of the sequential steps in the mitigation hierarchy, fair distribution changes along the different phases of wind-power development (Table 1). Even though the developer is the one that initiates the risk of harm, it does so because society desires more renewable energy (beneficiary of development). Local communities may both benefit from (e.g. investment, employment) and be affected by (e.g. visibility, noise) wind-power development, also giving them a dual role ethically speaking. Even non-governmental environmental organizations have an ethical role in the matter because they have taken responsibility for wildlife protection. Both local communities and non-governmental environmental organizations may therefore benefit from mitigation. R&D institutions are in principle non-beneficiaries, contributing with knowledge and tools, but there may exist financial dependencies making them in part a beneficiary of development. Lastly, the international community obtains neither benefits from the development and the mitigation (non-beneficiary) but may, for instance, in the case of wind-power development in developing countries, contribute to mitigation because of its ability to pay or set standards. All stakeholders have a certain role and responsibility for mitigation during the life cycle of a wind-power plant, even turbine companies, to produce wildlife-friendly wind turbines (Minimize, Table 1). Costs for mitigation therefore need to be fairly distributed relative to their responsibility versus need, in turn determining their contribution and capacity for implementing mitigation [68, 73]. Clearly, this is a difficult exercise due to mistrust and institutional settings [72, 74]. Active stakeholder participation and transparency are key to overcome such barriers through conservation triage [39] or adaptive management approaches [3, 75].

4 Implementing Mitigation

4.1 Trouble Never Comes Alone

If we only had to consider one species at any given time, with a clear impact and little uncertainty, then it would be relatively easy to deal with the ethical implications of mitigation. However, there are usually multiple species at risk, at multiple sites. Different species typically have varied ecological requirements, demographic characteristics, behaviour and habitat preferences. This complicates decision-making as each species (and site) may require different mitigation approaches [76]. Assessing potential risks of impact across species and ecosystem processes [17, 19] requires a good understanding of the system [12, 23, 77]. This is especially important because the species that receives most interest are not necessarily those that require it most (see Textbox). Similarly, the type of impact in focus may not have the largest ecological significance [78, 79]. Some ecological impacts may be enigmatic in nature, obscured from scrutiny for a variety of reasons [6]. Raiter and Possingham [6] identified four categories of enigmatic impacts: cumulative, off-site, cryptic and secondary impacts. While in certain circumstances an impact may be more obvious, in other contexts it may be overlooked. For instance, although the number of collisions is often limited within a single wind-power plant, cumulatively it may affect population perseverance across a species' geographic distribution [16, 80]. Avoidance enables wildlife to respond to the disturbance caused by the presence of wind turbines [13] but may however also lead to physiological stress responses [81] and ultimately affect populations [79, 82]. It may therefore be worthwhile to focus on key components of populations or ecosystems when managing wind-turbine-induced impacts on wildlife [23]. Identifying which species are expected to be vulnerable, and through which demographic characteristics [83], can direct interventions where they are needed most both pre- and post-construction.

Table 1 Responsibility burden shares for mitigation

Stakeholder	Principle role	Avoid	Minimize	Reduce	Compensate	Restore
Developer, operator	Risk initiator	Corporate responsibility	Polluter pays principle	Polluter pays principle	Polluter pays principle	Short-term 'kick-start'
Authorities, society at large	Beneficiary of development	Laws and regulations	Incentives, requirements	Incentives, requirements	Laws and regulations	Long-term maintenance
Local community	Beneficiary of development and mitigation	Mobilizing local knowledge	Facilitation, adaptation	Facilitation, adaptation	Mobilizing local knowledge	Long-term maintenance
Non-governmental organizations	Beneficiary of mitigation[a]	Disclosing species information	Disclosing species information	Conservation action	Conservation action	Conservation action
International community	Non-beneficiary	International conventions	International standards	International standards	Development aid	Development aid
R&D institutions	Non-beneficiary[b]	Tool development, empirical knowledge	Functional measures, monitoring	Functional measures, monitoring	Best practice methods, monitoring	Best practice methods, monitoring

Each row indicates which principle role each stakeholder holds and which actions they should entertain for each of the consecutive steps of the mitigation hierarchy

[a]Non-governmental organizations can also be a beneficiary of development with respect to wind power contributing to the reduction in carbon emissions, which in turn may have a positive effect on wildlife

[b]Financial dependencies on the developer/operator may exist, making the R&D institutions partly a beneficiary of development

The Smøla Case: Multiple Sides to Impacts and Mitigation

The Smøla wind-power plant has received a dubious reputation internationally because of the white-tailed eagles (*Haliaeetus albicilla*) that collide with the wind turbines. So, why did it come so far? The obligatory Environmental Impact Assessment at the start of the planning process suggested already then that a conflict with white-tailed might occur [84]. The Directorate for Nature Management as hearing partner recommended the need for mitigation measures to be implemented. Still, the wind-power plant received consent in 2000 taking all factors into consideration (e.g. Smøla has very good wind conditions, it would generate local income). Even though the planning process was characterized by its relative openness [85], there was local resistance to the construction of the wind-power plant, with the potential eagle conflict quickly adopted as its core argument [85, 86]. In 2001, Birdlife Norway filed a complaint to the Bern Convention regarding the impact the wind-power plant would have on the local white-tailed eagle population [87]. It was concluded however that normal procedures were followed during the planning phase both by the developer Statkraft and the consenting authority, the Directorate for Energy and Water Resources (NVE), but that procedures could be improved [87]. At the time of consenting, much less knowledge was available on the likelihood or seriousness of the anticipated harm on white-tailed eagles. The only comparable knowledge at that time was from Altamont Pass [88]. It was argued by NVE that construction would be proportional, because this wind-power plant would render much electricity and that the dense local eagle population could sustain a loss. After construction, the white-tailed eagle collisions triggered research on understanding the impacts of wind turbines on birds [89]. Since then much more knowledge on the impacts on white-tailed eagles has been obtained at Smøla [90], and several mitigation measures were tested in situ [e.g. painting rotor blades and tower bases, micrositing; 26]. This has reduced the uncertainty regarding both the extent of the impact and decision-making (e.g. the proposed wind-power plant on the neighbouring island of Frøya was temporarily stopped and later adjusted based on the research at Smøla). However, not only white-tailed eagles collide with the wind turbines, in fact the most common collision victim is the willow ptarmigan (*Lagopus lagopus*). The attention that the eagles attracted triggered research on impacts and potential mitigation measures for ptarmigans. Upon finalizing the research project testing mitigation measures, Statkraft obtained permanent exemption to leave the painted tower bases and rotor blades intact. This research effort has contributed to reducing the scientific uncertainty regarding the extent of impacts and proved the effectiveness of mitigation measures and thereby includes elements of adaptive management [75].

4.2 Embracing Uncertainty

The complexity of the social-ecological systems within which mitigation is set requires decision-makers to deal with different sources of uncertainty: natural stochasticity, statistical variance, social doubt, technical inefficiency and decision fickleness. Still, we need to work with what we know, and we must act to the best of our ability. The development history at Smøla shows elements of this (see Textbox). While the planning was relatively open, stakeholders were not held responsible for their role. At that time, scientific knowledge on the predicted impact was scant, and information of efficacy of mitigation measures was generally lacking. Consequently, consenting authorities were forced to decide in the face of uncertainty, which also affected the consenting process on the neighbouring island Frøya. This requires that we assess uncertainty of impacts across spatial and temporal scales and verify the efficacy of mitigation measures to make risk-based trade-offs in decision-making. It is unlikely that such trade-offs can be solely based on ecological premises but should rather encompass socioecological thresholds [91, 92]. As Loder [67] stated: 'Intolerance for uncertainty and risk comes at an ethical cost and may well impede implementation of policy goals'. Combining ethical theories to embrace uncertainty may well be our best bet to facilitate sustainable development. Combining utilitarian optimization with deontological proactive rules and temperance and care virtues may give us the opportunity for more robust decision-making in the face of uncertainty [39]. This entails trading some optimal performance (i.e. energy production) for less sensitivity to assumptions, satisficing over a wide range of futures and keeping options open [93]. This is especially important as mitigation generally takes effect in the future; precautions thus incur present costs to mitigate a future risk [21, 48]. For instance, although the common buzzard (*Buteo buteo*) is currently common in northern Germany, fatality studies showed that the species often collides with wind turbines and that the population was predicted to decline in 14 years due to this additional mortality [94]. Still, a large impact (cost) realized far into the future, becomes negligible in the present due to discounting [68, 73]. Similarly, time-lags in impacts (or knowledge on these) may support postponing mitigation. An implication of this is that stakeholder-relative reasons of the present make it lucrative to postpone mitigation costs as future generations are assumed to be richer. To avoid such a burden shift to future generations, adaptive management approaches will need to deal with intergenerational impartiality [21].

4.3 Managing Mitigation

As I hope I have been able to show in the above, managing mitigation is a multi-facetted decision-making challenge. Obtaining more knowledge on the probability and extent of a potential impact will not automatically solve the problem. It also hinges on how we as society perceive these impacts and how we prioritize

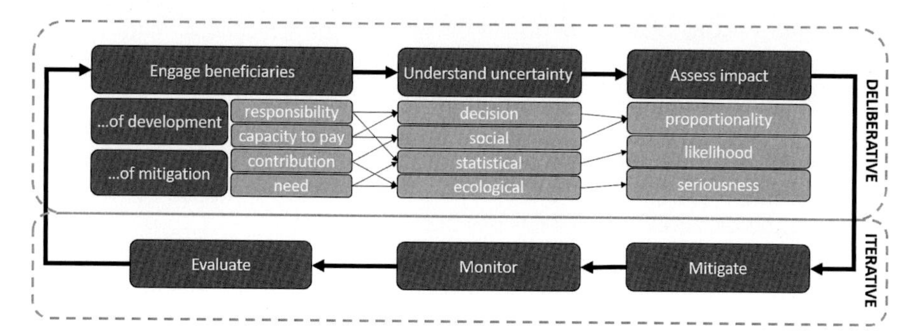

Fig. 2 Ethical considerations set within the deliberative planning phase of adaptive management framework

development relative to these. Conservation triage (i.e. the process of making difficult decisions regarding priorityunder severely constrained resources [39]) should encompass these spheres of interest (cf. Fig. 1) and needs therefore to be based on a more diverse set of ethics [9, 49]. This may require that we reconsider and restructure decision-making processes to be more transparent and inclusive, enabling social learning [9, 95]. Learning by doing includes using structured decision processes that make time scales of actions and consequences explicit [96], identify appropriate spatial scales for anticipatory predictions [97] and include participation to build trust among the stakeholders (cf. Table 1). In Schleswig-Holstein, Germany, participatory bottom-up governance worked well for locating new areas for wind power [74]. At the project-level, however, Statkraft invited stakeholders early in the planning process, but this did not reduce conflicts at the Smøla wind-power plant (see Textbox). This was likely due to strong local opposing views on the social impacts of the wind-power plant (i.e. turning negative economic trends versus compensating lost sociocultural values), which thereafter was transformed into a more technical issue of the extent of environmental risk [85]. Although the current paradigm is to deal with wind-turbine-induced impacts on wildlife based on 'return on investment', or in other words we mitigate only when we can be, certain measures are functional and will solve the problem in a technical sense. Given the current state of knowledge, we can however not wait until we know with certainty which mitigation measures are truly functional across species and sites. Instead we should adaptively manage impacts through learning by doing [3, 75]. The iterative loop Plan-Implement-Monitor-Evaluate of adaptive management has been promoted to help decision-makers balance development and conservation needs [98]. Williams and Brown [98] proposed a two-phase process, including a deliberative and iterative phase, for social and technical learning. The three central ethical questions discussed above may guide the deliberative phase to plan for mitigation (Fig. 2). Engagement of the beneficiaries of the development (including the risk initiator) and the mitigation (or even non-beneficiaries), and acknowledge their potential role in the adaptive process, can trigger trust among stakeholders [74]. Acknowledging a beneficiary's role within the decision process further provides insight into possible causes or origins of uncertainty perceptions. This in turn provides a better basis for assessing the extent of an impact and whether mitigation should be implemented. Improved deliberative planning will in turn enhance the iterative process of implementing, monitoring and evaluating management actions. Even though this may still entail making mistakes, it will enhance our collective knowledge base and build trust, gradually improving decision-making processes. By implementing those mitigation measures that are deemed best in a specific project and monitoring their effect in reducing impacts, as well as accepting a continuation of an impact in case it fails, we can however learn for future developments. Such a programme-based adaptive management approach [75] places the shared burden on all beneficiaries (cf. Table 1): mitigate locally but manage regionally.

Acknowledgements I would like to thank Espen Dyrnes Stabell and Daniel Steel for their valuable input during discussion on how I could apply their ethical framework for distributive fairness to the mitigation hierarchy for wind power. I also thank two anonymous reviewers for their critical comments that improved the contents of this chapter.

References

1. Gartman, V., Bulling, L., Dahmen, M., Geißler, G., Köppel, J.: Mitigation measures for wildlife in wind energy development, consolidating the state of knowledge — part 2: operation, decommissioning. JEAPM. **18**, 1650014 (2016). https://doi.org/10.1142/s1464333216500149
2. Vaissière, A.C., Levrel, H., Pioch, S., Carlier, A.: Biodiversity offsets for offshore wind farm projects: the current situation in Europe. Mar. Policy. **48**, 172–183 (2014). https://doi.org/10.1016/j.marpol.2014.03.023
3. Köppel, J., Dahmen, M., Helfrich, J., Schuster, E., Bulling, L.: Cautious but committed: moving toward adaptive planning and operation strategies for renewable energy's wildlife implications. Environ. Manag. **54**, 744–755 (2014). https://doi.org/10.1007/s00267-014-0333-8
4. Lovich, J.E., Ennen, J.R.: Assessing the state of knowledge of utility-scale wind energy development and operation on non-volant terrestrial and marine wildlife. Appl. Energy. **103**, 52–60 (2013). https://doi.org/10.1016/j.apenergy.2012.10.001
5. Martin, J.-L., Maris, V., Simberloff, D.S.: The need to respect nature and its limits challenges society and conservation science. Proc. Acad. Nat. Sci. Phila. **113**, 6105–6112 (2016). https://doi.org/10.1073/pnas.1525003113
6. Raiter, K.G., Possingham, H.P., Prober, S.M., Hobbs, R.J.: Under the radar: mitigating enigmatic ecological impacts. Trends Ecol. Evol. **29**, 635–644 (2014). https://doi.org/10.1016/j.tree.2014.09.003
7. Hayes, D.J.: Addressing the environmental impacts of large infrastructure projects: making "mitigation" matter. Environ. Law Report. **1**, 10016–10021 (2014)
8. Cole, S.G.: Wind power compensation is not for the birds: an opinion from an environmental economist. Restor. Ecol. **19**, 147–153 (2011). https://doi.org/10.1111/j.1526-100X.2010.00771.x
9. Sovacool, B.K., Heffron, R.J., McCauley, D., Goldthau, A.: Energy decisions reframed as justice and ethical concerns. Nat. Energy. **1**, (2016). https://doi.org/10.1038/nenergy.2016.24
10. Tallis, H., Kennedy, C.M., Ruckelshaus, M., Goldstein, J., Kiesecker, J.M.: Mitigation for one & all: an integrated framework for mitigation of development impacts on biodiversity and ecosystem services. Environ. Impact Assess. Rev. **55**, 21–34 (2015). https://doi.org/10.1016/j.eiar.2015.06.005

11. Künneke, R., Mehos, D.C., Hillerbrand, R., Hemmes, K.: Understanding values embedded in offshore wind energy systems: toward a purposeful institutional and technological design. Environ. Sci. Pol. **53**, 118–129 (2015). https://doi.org/10.1016/j.envsci.2015.06.013
12. May, R., Gill, A.B., Köppel, J., Langston, R.H.W., Reichenbach, M., Scheidat, M., Smallwood, S., Voigt, C.C., Hüppop, O., Portman, M.: Future research directions to reconcile wind turbine–wildlife interactions. In: Köppel, J., (ed.) Wind Energy and Wildlife Interactions: Presentations from the CWW2015 Conference, pp. 255–276. Springer, Cham (2017)
13. May, R.F.: A unifying framework for the underlying mechanisms of avian avoidance of wind turbines. Biol. Conserv. **190**, 179–187 (2015). https://doi.org/10.1016/j.biocon.2015.06.004
14. Anonymous.: Utility company sentenced in Wyoming for killing protected birds at wind projects. Justice News 2014 15.09.2014 [cited 2015 06.05.2015]; Available from: http://www.justice.gov/opa/pr/utility-company-sentenced-wyoming-killing-protected-birds-wind-projects
15. European Union: Guidance Document. Wind Energy Developments and Natura 2000. Publications Office of the European Union, Luxembourg (2011)
16. May, R., Masden, E.A., Bennet, F., Perron, M.: Considerations for upscaling individual effects of wind energy development towards population-level impacts on wildlife. J. Environ. Manage. **230**, 84–93 (2018). https://doi.org/10.1016/j.jenvman.2018.09.062
17. Gill, A.B.: Offshore renewable energy: ecological implications of generating electricity in the coastal zone. J. Appl. Ecol. **42**, 605–615 (2005). https://doi.org/10.1111/j.1365-2664.2005.01060.x
18. Valiente-Banuet, A., Aizen, M.A., Alcántara, J.M., Arroyo, J., Cocucci, A., Galetti, M., García, M., García, D., Gómez, J., Jordano, P., Medel, R., Navarro, L., Obeso, J.R., Oviedo, R., Ramírez, N., Traveset, A., Verdú, M., Zamora, R.: Beyond species loss: the extinction of ecological interactions in a changing world. Funct. Ecol. **29**, 299–307 (2015). https://doi.org/10.1111/1365-2435.12356
19. Burkhard, B., Opitz, S., Lenhart, H., Ahrendt, K., Garthe, S., Mendel, B., Windhorst, W.: Ecosystem based modeling and indication of ecological integrity in the German North Sea-Case study offshore wind parks. Ecol. Indic. **11**, 168–174 (2011). https://doi.org/10.1016/j.ecolind.2009.07.004
20. Cardona, O.-D., van Aalst, M.K., Birkmann, J., Fordham, M., McGregor, G., Perez, R., Pulwarty, R.S., Schipper, E.L.F., Sinh, B.T.: Determinants of risk: exposure and vulnerability. In: Field, C.B. et al. (eds.) Managing the Risks of Extreme Events and Disasters to Advance Climate Change Adaptation. A Special Report of Working Groups I and II of the Intergovernmental Panel on Climate Change (IPCC), pp. 65–108. Cambridge University Press, Cambridge
21. Steel, D.: Philosophy and the Precautionary Principle. Science, Evidence, and Environmental Policy. Cambridge University Press, Cambridge (2015)
22. von Schomberg, R.: The precautionary principle and its normative challenges. In: Fisher, E., Jones, J., von Schomberg, R. (eds.) Implementing the Precautionary Principle: Perspectives and Prospects, pp. 19–42. Edward Elgar, Cheltenham/Northampton (2006)
23. Hobbs, R.J., Hallet, L.M., Ehrlich, P.R., Mooney, H.A.: Intervention ecology: applying ecological science in the twenty-first century. Bioscience. **61**, 442–450 (2011). https://doi.org/10.1525/bio.2011.61.6.6
24. Wiens, J.A., Hobbs, R.J.: Integrating conservation and restoration in a changing world. Bioscience. **65**, 302–312 (2015). https://doi.org/10.1093/biosci/biu235
25. Luuppala, L.S.: Ecological Restoration: Conceptual Analysis and Ethical Implications. University of Helsinki. Helsinki, Finland (2015)
26. May, R.: Mitigation options for birds. In: Perrow, M. (ed.) Wildlife and Windfarms: Conflicts and Solutions Onshore Solutions, vol. 2, pp. 124–145. Pelagic Publishing, Exeter (2017)
27. McGinnis, M.D., Ostrom, E.: Social-ecological system framework: initial changes and continuing challenges. Ecol. Soc. **19**, (2014). Artn 30). https://doi.org/10.5751/Es-06387-190230
28. Ostrom, E.: A general framework for analyzing sustainability of social-ecological systems. Science. **325**, 419–422 (2009). https://doi.org/10.1126/science.1172133

29. Smeets, E., Weterings, R.: Environmental Indicators: Typology and Overview. E.E. Agency, Copenhagen (1999)
30. EPA: Guidelines for Ecological Risk Assessment. Environmental Protection Agency, Washington, DC (1998)
31. Attfield, R.: Environmental Ethics: An Overview. eLS, A24201 (2012)
32. Schaub, M.: Spatial distribution of wind turbines is crucial for the survival of red kite populations. Biol. Conserv. **155**, 111–118 (2012). https://doi.org/10.1016/j.biocon.2012.06.021
33. Klain, S.C., Satterfield, T., Sinner, J., Ellis, J.I., Chan, K.M.A.: Bird killer, industrial intruder or clean energy? Perceiving risks to ecosystem services due to an offshore wind farm. Ecol. Econ. **143**, 111–129 (2018). https://doi.org/10.1016/j.ecolecon.2017.06.030
34. Menegaki, A.: Valuation for renewable energy: a comparative review. Renew. Sustain. Energy Rev. **12**, 2422–2437 (2008). https://doi.org/10.1016/j.rser.2007.06.003
35. Welstead, J., Hirst, R., Keogh, D., Robb, G., Bainsfair, R.: Research and Guidance on Restoration and Decommissioning of Onshore Wind Farms. Scottish Natural Heritage, Inverness (2013)
36. Aydin, N.Y., Kentel, E., Duzgun, S.: GIS-based environmental assessment of wind energy systems for spatial planning: a case study from Western Turkey. Renew. Sustain. Energy Rev. **14**, 364–373 (2010). https://doi.org/10.1016/j.rser.2009.07.023
37. Tsoutsos, T., Tsitoura, I., Kokologos, D., Kalaitzakis, K.: Sustainable siting process in large wind farms case study in Crete. Renew. Energy. **75**, 474–480 (2015). https://doi.org/10.1016/j.renene.2014.10.020
38. Ives, C.D., Bekessy, S.A.: The ethics of offsetting nature. Front. Ecol. Evol. **13**, 568–573 (2015). https://doi.org/10.1890/150021
39. Wilson, K.A., Law, E.A.: Ethics of conservation triage. Front. Ecol. Evol. **4**, 112 (2016). https://doi.org/10.3389/fevo.2016.00112
40. Gove, B., Langston, R.H.W., McCluskie, A., Pullan, J.D., Scrase, I.: Wind Farms and Birds: An Updated Analysis of the Effects of Wind Farms on Birds, and Best Practice Guidance on Integrated Planning and Impact Assessment. C.o. Europe, Strasbourg (2013)
41. IFC: Environmental, Health, and Safety Guidelines Wind Energy. I.F. Corporation (2015)
42. U.S. Fish & Wildlife Service: U.S. Fish and Wildlife Service Land-Based Wind Energy Guidelines. Arlington (2012)
43. Hanssen, F., May, R., van Dijk, J., Stokke, B.G., De Stefano, M.: Spatial Multi-Criteria Decision Analysis (SMCDA) Toolbox for Consensus-Based Siting of Powerlines and Wind-Power Plants (Con-Site). Norwegian Institute for Nature Research, Trondheim (2018)
44. Dahl, E.L., May, R., Nygård, T., Åstrøm, J., Diserud, O.H.: Repowering Smøla Wind Power Plant. An Assessment of Avian Conflicts. Norwegian Institute for Nature Research, Trondheim (2015)
45. Arnett, E.B., Huso, M.M.P., Schirmacher, M.R., Hayes, J.P.: Altering turbine speed reduces bat mortality at wind-energy facilities. Front. Ecol. Environ. **9**, 209–214 (2011). https://doi.org/10.1890/100103
46. Hendersen, N., Sutherland, W.J.: Two truths about discounting and their environmental consequences. Trends Ecol. Evol. **11**, 527–528 (1996)
47. Quetier, F., Lavorel, S.: Assessing ecological equivalence in biodiversity offset schemes: key issues and solutions. Biol. Conserv. **144**, 2991–2999 (2011). https://doi.org/10.1016/j.biocon.2011.09.002
48. Moilanen, A., van Teeffelen, A.J.A., Ben-Haim, Y., Ferrier, S.: How much compensation is enough? A framework for incorporating uncertainty and time discounting when calculating offset ratios for impacted habitat. Restor. Ecol. **17**, 470–478 (2009). https://doi.org/10.1111/j.1526-100X.2008.00382.x
49. Jenkins, K., McCauley, D., Heffron, R., Stephan, H., Rehner, R.: Energy justice: a conceptual review. Energy Res. Soc. Sci. **11**, 174–182 (2016). https://doi.org/10.1016/j.erss.2015.10.004
50. Duinker, P.N.: FORUM: the significance of environmental impacts: an exploration of the concept. Environ. Manag. **10**, 1–10 (1986)

51. Lawrence, D.P.: Impact significance determination—back to basics. Environ. Impact Assess. Rev. **27**, 755–769 (2007). https://doi.org/10.1016/j.eiar.2007.02.011
52. Baerwald, E.F., Edworthy, J., Holder, M., Barclay, R.M.R.: A large-scale mitigation experiment to reduce bat fatalities at wind energy facilities. J. Wildl. Manag. **73**, 1077–1081 (2009). https://doi.org/10.2193/2008-233
53. Arnett, E.B., M. Baker, C. Hein, M. Schirmacher, M.M.P. Huso, J.M. Szewczak: Effectiveness of deterrents to reduce bat fatalities at wind energy fatalities. Proceedings Conference on Wind Energy and Wildlife impacts, 2–5 May 2011, Trondheim, Norway. NINA Report 693, R. Bevanger K, May, R 57. Norwegian Institute for Nature Research, Trondheim (2011)
54. Dahl, E.L.: Population Demographics in White-Tailed Eagle at an On-Shore Wind Farm Area in Coastal Norway. Norwegian University of Science and Technology (NTNU), Trondheim (2014)
55. Trouwborst, A.: Prevention, precaution, logic and law. The relationship between the precautionary principle and the preventative principle in international law and associated questions. Erasmus Law Rev. **2**, 105–127 (2009)
56. Gardiner, S.M., Core Precautionary, A.: Principle. J. Polit. Philos. **14**, 33–60 (2006)
57. Hermerén, G.: The principle of proportionality revisited: interpretations and applications. Med. Health Care Philos. **15**, 373–382 (2012). https://doi.org/10.1007/s11019-011-9360-x
58. Gaspars-Wieloch, H.: Modifications of the Hurwicz's decision rule. CEJOR. **22**, 779–794 (2014). https://doi.org/10.1007/s10100-013-0302-y
59. May, R., Reitan, O., Bevanger, K., Lorentsen, S.H., Nygard, T.: Mitigating wind-turbine induced avian mortality: sensory, aerodynamic and cognitive constraints and options. Renew. Sust. Energ. Rev. **42**, 170–181 (2015). https://doi.org/10.1016/j.rser.2014.10.002
60. Marques, A.T., Batalha, H., Rodrigues, S., Costa, H., Pereira, M.J.R., Fonseca, C., Mascarenhas, M., Bernardino, J.: Understanding bird collisions at wind farms: an updated review on the causes and possible mitigation strategies. Biol. Conserv. **179**, 40–52 (2014). https://doi.org/10.1016/j.biocon.2014.08.017
61. Gartman, V., Bulling, L., Dahmen, M., Geißler, G., Köppel, J.: Mitigation measures for wildlife in wind energy development, consolidating the state of knowledge—part 1: planning and siting, construction. JEAPM. **18**, 1650013 (2016). https://doi.org/10.1142/s1464333216500137
62. Gardner, T.A., von Hase, A., Brownlie, S., Ekstrom, J.M., Pilgrim, J.D., Savy, C.E., Stephens, R.T., Treweek, J., Ussher, G.T., Ward, G., Ten Kate, K.: Biodiversity offsets and the challenge of achieving no net loss. Conserv. Biol. **27**, 1254–1264 (2013). https://doi.org/10.1111/cobi.12118
63. IFC: Performance Standard 6. I.F. Corporation (2012)
64. Langhamer, O.: Artificial reef effect in relation to offshore renewable energy conversion: state of the art. Sci. World J. **2012**, 386713 (2012). https://doi.org/10.1100/2012/386713
65. Smyth, K., Christie, N., Burdon, D., Atkins, J.P., Barnes, R., Elliott, M.: Renewables-to-reefs? – decommissioning options for the offshore wind power industry. Mar. Pollut. Bull. **90**, 247–258 (2015). https://doi.org/10.1016/j.marpolbul.2014.10.045
66. Virah-Sawmy, M., Ebeling, J., Taplin, R.: Mining and biodiversity offsets: a transparent and science-based approach to measure "no-net-loss". J. Environ. Manag. **143**, 161–170 (2014). https://doi.org/10.1016/j.jenvman.2014.03.027
67. Loder, R.E.: Breath of life: ethical wind power and wildlife. Vermont J. Environ. Law. **10**, 507–531 (2009)
68. Stabell, E.D., Steel, D.: Precaution and fairness: a framework for distributing costs of protection from environmental risks. J. Agric. Environ. Ethics. (2018). https://doi.org/10.1007/s10806-018-9709-8
69. Aggarwal, R., Dow, S.: Corporate governance and business strategies for climate change and environment mitigation. Eur. J. Financ. **18**, 113–131 (2012)
70. Unsworth, K.L., Russell, S.V., Davis, M.C.: Is dealing with climate change a corporation's responsibility? A social contract perspective. Front. Psychol. **7**(1212), (2016). https://doi.org/10.3389/fpsyg.2016.01212

71. Warren, C.R., Lumsden, C., O'Dowd, S., Birnie, R.V.: 'Green on green': public perceptions of wind power in Scotland and Ireland. J. Environ. Plan. Manag. **48**, 853–875 (2005)
72. Wolsink, M.: Wind power: basic challenge concerning social acceptance. In: Meyers, R.A. (ed.) Encyclopedia of Sustainability Science and Technology, pp. 12218–12254. Springer-Verlag, New York (2012)
73. Ringius, L., Torvanger, A., Underdal, A.: Burden sharing and fairness principles in international climate policy. Int. Environ. Agreements: Polit. Law Econ. **2**, 1–22 (2002)
74. Gartman, V., Wichmann, K., Bulling, L., Elena Huesca-Perez, M., Koppel, J.: Wind of change or wind of challenges: implementation factors regarding wind energy development, an international perspective. AIMS Energy. **2**, 485–504 (2014). https://doi.org/10.3934/energy.2014.4.485
75. Hanna, L., A. Copping, S. Geerlofs, L. Feinberg, J. Brown-Saracino, F. Bennett, R. May, J. Köppel, L. Bulling, V. Gartman: Results of IEA Wind Adaptive Management White Paper. Prepared for the International Energy Agency Wind Implementing Agreement. I.E.A. Wind (2016)
76. Arnett, E.B., May, R.F.: Mitigating wind energy impacts on wildlife: approaches for multiple taxa. Hum. Wildl. Interact. **10**, 28–41 (2016)
77. Mattews, H.D., Turner, S.E.: Of mongooses and mitigation: ecological analogues to geoengineering. Environ. Res. Lett. **4**, 045105 (2009). https://doi.org/10.1088/1748-9326/4/4/045105
78. Gill, J.A., Norris, K., Sutherland, W.J.: Why behavioural responses may not reflect the population consequences of human disturbance. Biol. Conserv. **97**, 265–268 (2001)
79. Frid, A., Dill, L.: Human-caused disturbance stimuli as a form of predation risk. Conserv. Ecol. **6**, 11 (2002)
80. Masden, E.A., McCluskie, A., Owen, E., Langston, R.H.W.: Renewable energy developments in an uncertain world: the case of offshore wind and birds in the UK. Mar. Policy. **51**, 169–172 (2015). https://doi.org/10.1016/j.marpol.2014.08.006
81. Agnew, R.C., Smith, V.J., Fowkes, R.C.: Wind turbines cause chronic stress in badgers (*Meles Meles*) in Great Britain. J. Wildl. Dis. **52**, 459–467 (2016). https://doi.org/10.7589/2015-09-231
82. King, S.L., Schick, R.S., Donovan, C., Booth, C.G., Burgman, M., Thomas, L., Harwood, J.: An interim framework for assessing the population consequences of disturbance. Methods Ecol. Evol. **6**, 1150–1158 (2015). https://doi.org/10.1111/2041-210x.12411
83. Thaxter, C.B., Buchanan, G.M., Carr, J., Butchart, S.H.M., Newbold, T., Green, R.E., Tobias, J.A., Foden, W.B., O'Brien, S., Pearce-Higgins, J.W.: Bird and bat species' global vulnerability to collision mortality at wind farms revealed through a trait-based assessment. Proc. Biol. Sci. **284**, 20170829 (2017). https://doi.org/10.1098/rspb.2017.0829
84. Follestad, A., Reitan, O., Pedersen, H.C., Brøseth, H., Bevanger, K.: Vindkraftverk på Smøla: Mulige konsekvenser for "rødlistede" fuglearter. N.I.f.N. Research, Trondheim (1999)
85. Solli, J.: Where the eagles dare? Enacting resistance to wind farms through hybrid collectives. Environ. Polit. **19**, 45–60 (2010). https://doi.org/10.1080/09644010903396077
86. Rygg, B.J.: Wind power—an assault on local landscapes or an opportunity for modernization? Energ Policy. **48**, 167–175 (2012). https://doi.org/10.1016/j.enpol.2012.05.004
87. Kuijken, E.: On-the-spot appraisal Wind farms at the Smøla Archipelago (Norway). 15–17 June 2009. Standing Committee of the Convention on the conservation of European wildlife and natural habitats, Strasbourg, France (2009)
88. Thelander, C.G., K.S. Smallwood: The altamont pass wind resource area's effects on birds: a case history. In: de Lucas, M., Janss, G.F.E., Ferrer, M. (eds.) Birds and Wind Farms. Risk Assessment and Mitigation, pp. 25–46. Servicios Informativos Ambientales/Quercus, Madrid (2007)
89. Bevanger, K., Berntsen, F., Clausen, S., Dahl, E.L., Flagstad, Ø., Follestad, A., Halley, D., Hanssen, F., Johnsen, L., Kvaløy, P., Lund-Hoel, P., May, R., Nygård, T., Pedersen, H.C., Reitan, O., Røskaft, E., Steinheim, Y., Stokke, B., Vang, R.: Pre- and Post-Construction Studies of Conflicts Between Birds and Wind Turbines in Coastal Norway (BirdWind). Report on findings 2007–2010. N.I.f.N. Research, Trondheim (2010)

90. Bevanger, K., R. May, B. Stokke: Landbasert vindkraft. Utfordringer for fugl, flaggermus og rein. N.I.f.N. Research, Trondheim, Norway (2016)
91. Cook, C.N., de Bie, K., Keith, D.A., Addison, P.F.E.: Decision triggers are a critical part of evidence-based conservation. Biol. Conserv. **195**, 46–51 (2016). https://doi.org/10.1016/j.biocon.2015.12.024
92. Martin, J., Runge, M.C., Nichols, J.D., Lubow, B.C., Kendall, W.L.: Structured decision making as a conceptual framework to identify thresholds for conservation and management. Ecol. Appl. **19**, 1079–1090 (2009). https://doi.org/10.1890/08-0255.1
93. Lempert, R.J., Collins, M.T.: Managing the risk of uncertain threshold responses: comparison of robust, optimum, and precautionary approaches. Risk Anal. **27**, 1009–1026 (2007). https://doi.org/10.1111/j.1539-6924.2007.00940.x
94. Grünkorn, T., Blew, J., Coppack, T., Krüger, O., Nehls, G., Potiek, A., Reichenbach, M., von Rönn, J., Timmermann, H., Weitekamp, S.: Prognosis and Assessment of Bird Collision Risks at Wind Turbines in Northern Germany (PROGRESS). Final report commissioned by the Federal Ministry for Economic affairs and Energy in the framework of the 6th Energy research programme of the federal government. BioConsult/ARSU/IfAÖ/University of Bielefeld, Husum/Oldenburg/Rostock/Bielefeld (2016)
95. Schlüter, M., Müller, B., Frank, K.: How to use models to improve analysis and governance of social-ecological systems – the reference frame MORE. SSRN (2013). https://doi.org/10.2139/ssrn.2037723
96. Wilson, R.S., Hardisty, D.J., Epanchin-Niell, R.S., Runge, M.C., Cottingham, K.L., Urban, D.L., Maguire, L.A., Hastings, A., Mumby, P.J., Peters, D.P.: A typology of time-scale mismatches and behavioral interventions to diagnose and solve conservation problems. Conserv. Biol. **30**, 42–49 (2016). https://doi.org/10.1111/cobi.12632
97. Mouquet, N., Lagadeuc, Y., Devictor, V., Doyen, L., Duputie, A., Eveillard, D., Faure, D., Garnier, E., Gimenez, O., Huneman, P., Jabot, F., Jarne, P., Joly, D., Julliard, R., Kefi, S., Kergoat, G.J., Lavorel, S., Le Gall, L., Meslin, L., Morand, S., Morin, X., Morlon, H., Pinay, G., Pradel, R., Schurr, F.M., Thuiller, W., Loreau, M.: REVIEW: predictive ecology in a changing world. J. Appl. Ecol. **52**, 1293–1310 (2015). https://doi.org/10.1111/1365-2664.12482
98. Williams, B.K., Brown, E.D.: Adaptive management: from more talk to real action. Environ. Manag. **53**, 465–479 (2014). https://doi.org/10.1007/s00267-013-0205-7

The First Large-Scale Offshore Aerial Survey Using a High-Resolution Camera System

Stephanie McGovern, Julia Robinson Wilmott, Gregory Lampman, Ann Pembroke, Simon Warford, Mark Rehfisch, and Stuart Clough

Abstract Aerial digital surveying techniques using aircraft flying at significantly higher and safer altitudes than observer-based aerial surveys have become a key tool in surveying offshore environments worldwide. In preparation for offshore wind energy development, the New York State Energy Research and Development Authority (NYSERDA) has initiated what is probably the world's largest (43,500 km^2) and highest-resolution offshore aerial survey of marine wildlife. The survey uses ultrahigh-resolution aerial digital imagery captured with the purpose-built camera system "Shearwater III" at 1.5 cm ground sampling distance (GSD) from a twin-engine aircraft flying at 415 m. This baseline study using an innovative survey technique will facilitate a more efficient planning of energy production offshore by providing the necessary information to meet regulatory requirements for environmental review of wind energy areas (WEAs). This review includes a high-level summary of methods and highlights the advantages of undertaking the NYSERDA surveys using high-resolution digital aerial techniques.

Keywords Digital aerial survey · Offshore bird surveys · New York state

S. McGovern (✉) · S. Warford · M. Rehfisch
APEM Ltd, Stockport, UK
e-mail: s.mcgovern@apemltd.co.uk; s.warford@apemltd.co.uk; m.rehfisch@apemltd.co.uk

J. R. Wilmott
Normandeau Associates Inc, Gainesville, FL, USA
e-mail: jwillmott@normandeau.com

G. Lampman
NYSERDA, Albany, NY, USA
e-mail: Gregory.Lampman@nyserda.ny.gov

A. Pembroke
Normandeau Associates Inc, Bedford, NH, USA
e-mail: apembroke@normandeau.com

S. Clough
APEM Inc, Gainesville, FL, USA
e-mail: s.clough@apem-inc.com

R. Bispo et al. (eds.), *Wind Energy and Wildlife Impacts*,
https://doi.org/10.1007/978-3-030-05520-2_7

1 Background to Avian Offshore Surveys

Large targets for offshore wind energy implementation have been set in recent years. For example, the European Union has set a target that 27% of its energy should be from renewables by 2030, and of that offshore wind will be a major contributor [1]. Like all large infrastructure projects, before an offshore wind farm (OWF) can be built, it must be preceded by an Environmental Impact Assessment (EIA) to determine whether the scale of its impact on the environment is acceptable [2]. As part of this EIA, the impact of the OWF on the wildlife in the proposed development site must be determined. As a first step towards this assessment, the birds on the site have to be surveyed in detail to provide a baseline of the species' abundance against which possible impacts can be predicted preconstruction and measured post-construction. The number and scale of the proposed OWF developments has led to the rapid development of new and improved technologies for the rapid and cost-efficient biological survey of the offshore environment.

Originally undertaken by boats and aerial visual surveys [3], technological developments have meant that more surveys are now being delivered by digital methods [4–6]. Whilst digital aerial surveys are suitable for many taxa, this review focuses on birds.

1.1 Visual Boat-Based Surveys

Boat-based surveys are a well-established technique for undertaking offshore bird surveys [3]. Boat-based surveys use a strip-transect methodology, essentially visual spotters viewing either side of the vessel as it moved through the sea. Standard methodologies, such as European Seabirds at Sea (ESAS), have been used extensively for this type of work. Boat-based surveys use two different methods for recording birds on the water and birds in the air. Birds on the water are recorded continuously with flying birds recorded in snapshot counts at set time intervals dependent on the speed of the vessel. These methods rely heavily on instantaneous identification of a species and counting of the numbers present. The data also need correcting during statistical analysis for the reduction in detectability as the distance from the survey vessel increases. Due to behavioural responses of individual species, bias can also be introduced into the survey data, as species can be either attracted to, or deterred, from the survey vessel [7].

1.2 Visual Aerial Surveys

Another well-established technique is visual aerial surveys. Visual aerial surveys are undertaken by an observer in a small aircraft, often flying along transects at around 80 m altitude, to record target birds [3]. Visual aerial surveys are appropriate for the rapid surveys of avian species over quite large areas [4]. They require observers that are highly trained in the identification of target species, with the ability to count accurately at speed, as there can be large, discrete multispecies aggregations of marine fauna [4]. The low flight altitude required for visual aerial surveys can cause disturbance to birds and can lead to a danger of accident by bird strike. Wilfully causing disturbance is not acceptable on Natura 2000 and many other designated sites. As with visual boat-based surveys, the data from visual aerial surveys require correcting for imperfect detectability during statistical analysis. With both visual survey techniques, it is not normally possible to undertake quality assurance on the data, check identifications or revisit the data at a later date as there is no photographic record of the data gathered [8].

1.3 Digital Aerial Surveys

Due to major advances in technology, digital aerial surveys are increasingly used for offshore wildlife surveys. Imagery is collected using specialised camera systems mounted on specially modified aircraft. The altitude at which digital surveys are flown is considerably higher than that of visual surveys which has the major advantage that wildlife are not disturbed [9, 10], thus complying with published recommendations [6].

During the early days of digital aerial surveys, image quality suffered due to aircraft vibration, and the resolution of the images made it difficult to identify smaller species. The development of gyrostabilised camera mounts for the aircraft overcame this issue, and technological developments have allowed greater resolution of images to aid in identifications [7].

Digital stills imagery provides a permanent record that provides the ability to return to the imagery at a later date and allows full quality control procedures to be implemented at every stage of the analysis. Numbers and identification of species can be checked by independent analysts providing a high level of confidence in the data.

A major strength of digital stills photography is that it can utilise different survey designs according to project requirements, such as employing either transect- or a grid-based survey designs. Surveys undertaken using the traditional visual methods were normally undertaken using transect-based designs. Transect surveys provide a cost-effective way to cover large areas relatively quickly but may not provide the spatial resolution required to answer more specific questions about the impact of developments on species numbers and distribution. Grid-based surveys, which collect a series of independent images across the survey area, are more time intensive for data collection for the same percentage coverage as a transect design but provide a finer scale of data, more evenly spaced throughout the survey area (Fig. 1), to allow BACI (before-after-change-impact) or BAG (before-after-gradient) effects to be detected. Digital aerial surveys can provide the necessary information required for Environmental Impact Assessment (EIA) such as flight heights for birds.

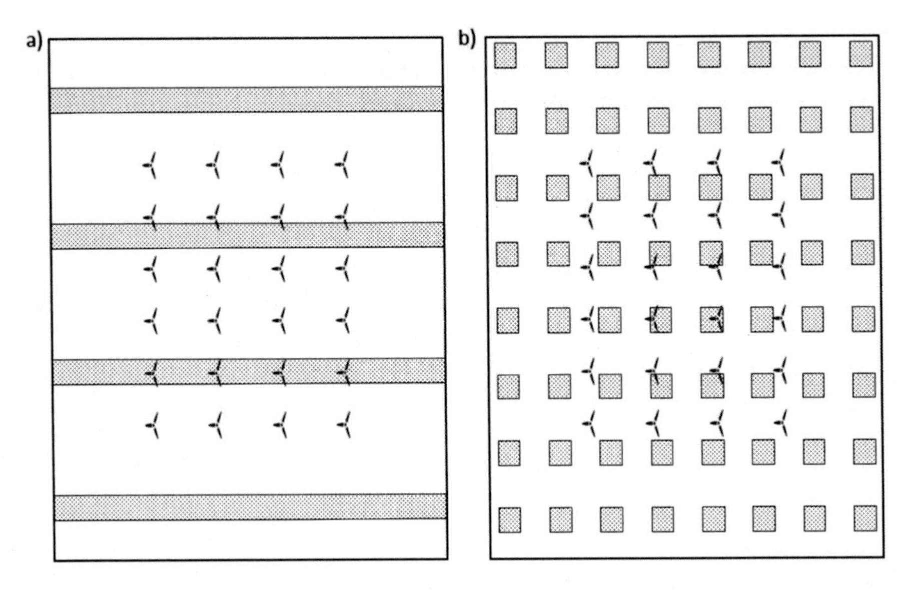

Fig. 1 Example of similar coverage of a wind farm and buffer area using (**a**) a transect-based approach and (**b**) a grid-based approach

On behalf of the Bureau of Ocean Energy Management, Normandeau Associates compared different methodologies for conducting offshore surveys [7]. The study's overarching conclusion is that high-resolution digital aerial imaging does represent a scientifically robust and cost-effective solution for offshore surveys, and this study further broke this down into three criteria as follows:

- Cost – high-resolution digital aerial imaging provides cost savings against conventional visual observer surveys.
- Safety – high-resolution digital aerial surveys are safer than visual observer-based aerial surveys due to higher flight altitudes.
- Effectiveness – high-resolution digital aerial imaging surveys have many advantages over conventional methods including:
  - Counts not distorted by the animals' attraction and displacement to the survey platform.
  - Survey swath can be calculated precisely from an image and is not subject to observers' error-prone distance estimations which allows for more precise estimates of density.
  - No observer swamping or search image effects.
  - Interobserver variability can be reduced through post hoc multiple observer quality assurance/quality control (QA/QC) image review.

Typically, digital aerial surveys completed to underpin environmental impact assessment are undertaken at 2 or 3 cm ground sampling distance (GSD). The GSD indicates the size of the pixel of the image at the sea surface. This allows for good coverage and high identification rates. However, where it is difficult to differentiate between key consenting species such as roseate, Arctic and common terns, higher resolutions may be necessary. Roseate tern is a species listed as threatened or endangered in a number of locations across the world. Inability to differentiate between a species of lesser conservation concern and a listed species could result in an offshore wind project placed at risk due to conservation concerns. Using a higher resolution, such as 1.5 cm, increases the identification rates obtained for very similar species, thus strengthening the assessment of impact for the species and by so doing reducing the level of consenting risk to a project. Positive identification of roseate terns has been achieved for this project with 1.5 cm resolution images. Digital survey data had a higher rate of identification to the species level (77%) compared with visual observer survey platforms (71% and 45% for boat and aerial visual observer surveys, respectively), when the methodologies were directly compared [7]. Example images at 3 cm, 2 cm and 1.5 cm are shown in Figs. 2 and 3. These images show the increased detail achieved with higher resolutions.

Digital stills surveys have the major strength that they can employ two different survey techniques, grid-based and transect-based surveys, and can undertake these much faster than traditional boat surveys. Previous studies [5] that investigated the difference between grid-based and transect-based surveys showed that for clumped species a more even survey coverage, such as that achieved by grids, provided more precise population estimates than a transect-based approach. However, for widely spaced pelagic species, there was less difference between the methods. Tailoring the survey design to ensure that they are suitable to the aim of the studies is essential. Even coverage with more independent replicates, such as those provided by a grid-based design, may provide more precise estimates [11] and allow BACI effects to be more readily determined. However to provide a general baseline of species present in a large area, transects are a more efficient means of collecting data.

2 Implementing the World's Largest Offshore High-Resolution Digital Aerial Survey

New York state has set a target to establish up to 2.4 GW of offshore wind by 2030 having adopted the "50 by '30" standard (50% of New York's electricity from truly renewable and pollution-free energy resources by 2030) [12]. In preparation for offshore wind energy development, the New York State Energy Research and Development Authority (NYSERDA) has initiated a large-scale offshore high-resolution digital aerial survey of marine wildlife spanning 3 years. Initial surveys were undertaken on both the offshore planning area (OPA) to provide a baseline dataset and a more focussed survey on a proposed wind energy area (WEA). These areas differ in size with the OPA covering a very large area of 43,500 km^2 and the WEA including a buffer covering an area of 850 km^2. These surveys use a

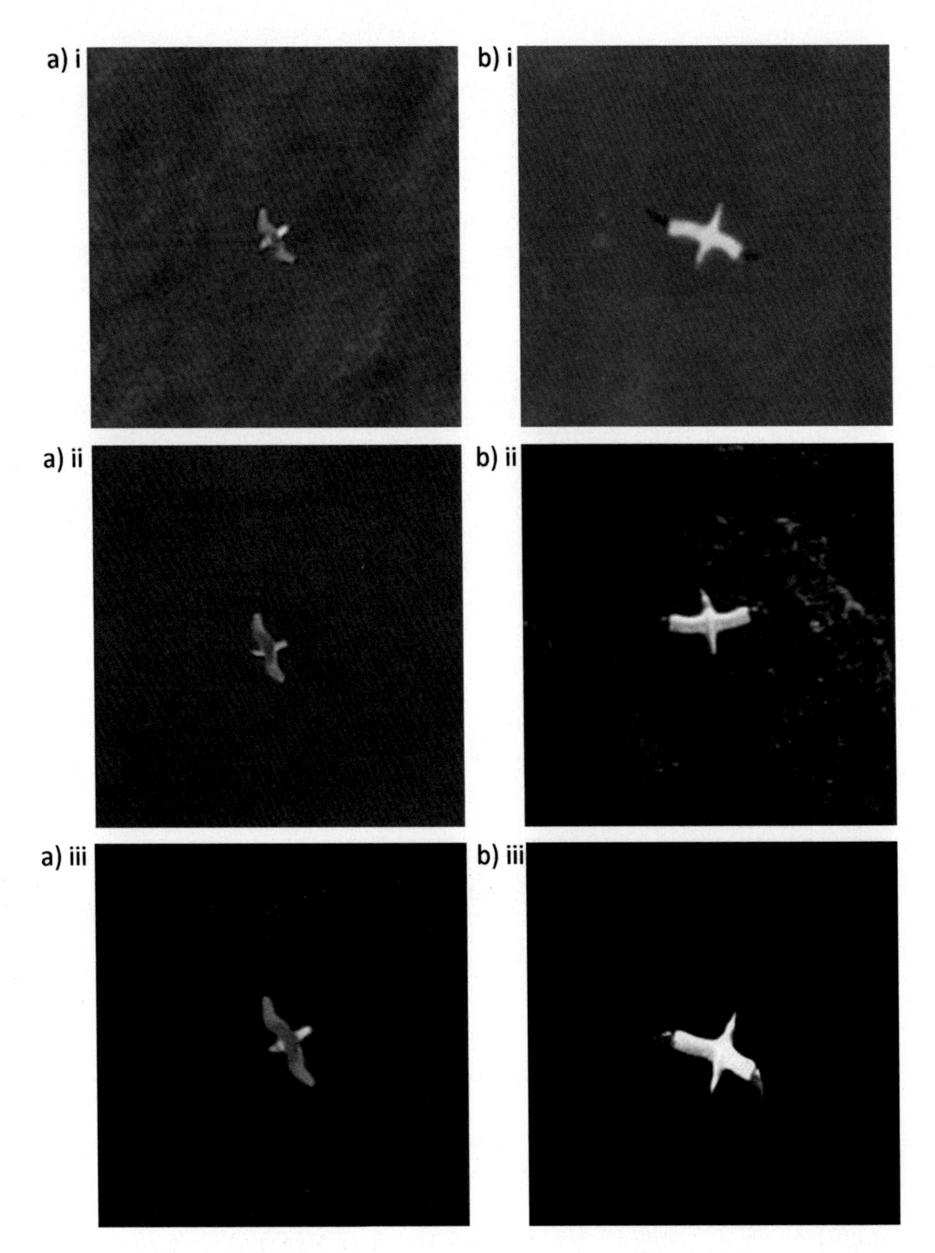

Fig. 2 Example images of (**a**) kittiwakes and (**b**) gannets at (i) 3 cm resolution, (ii) 2 cm resolution and (iii) 1.5 cm resolution

newly developed purpose-built camera system "Shearwater III" to capture ultrahigh-resolution aerial digital imagery at 1.5 cm GSD from a twin-engine aircraft flying at 415 m.

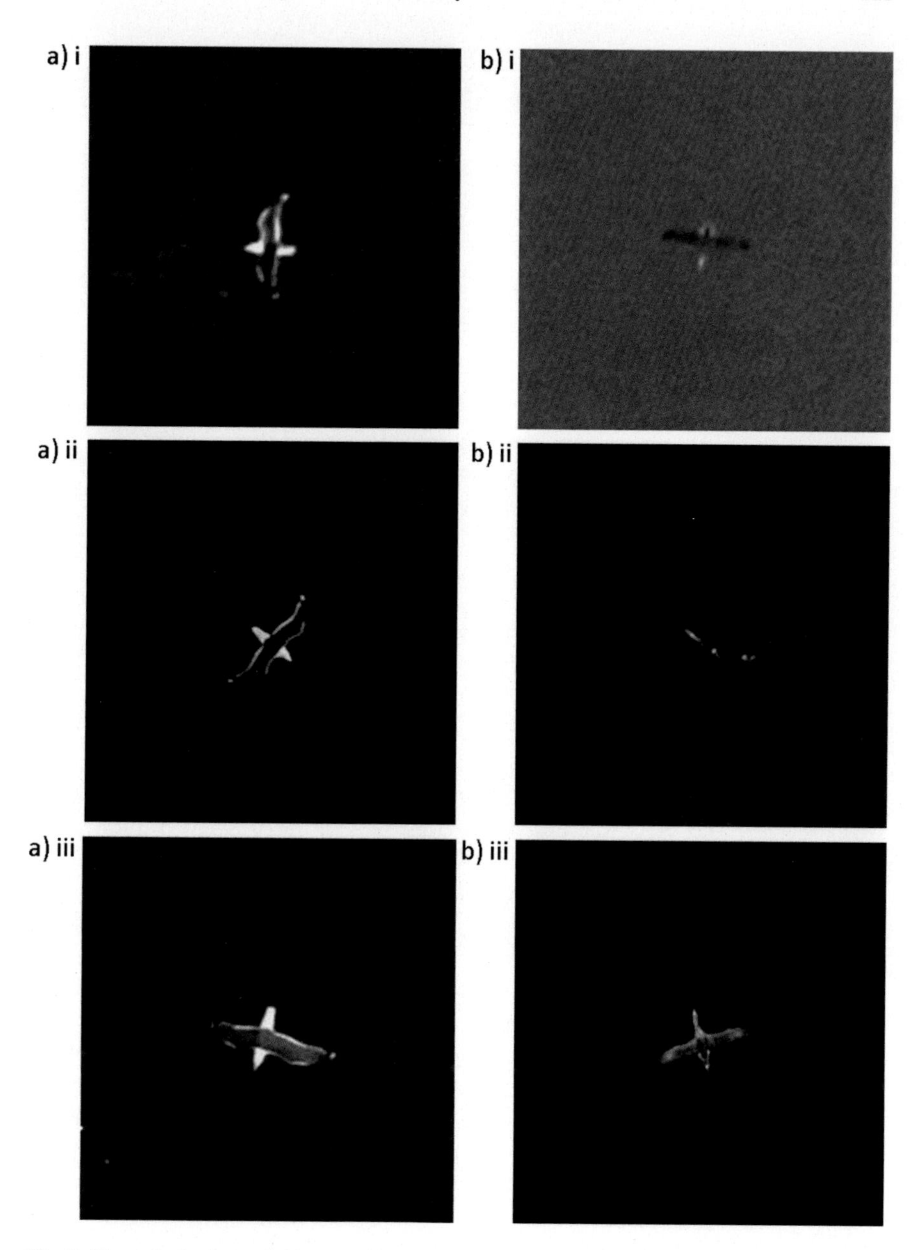

Fig. 3 Example images of (**a**) great black-backed gull and (**b**) red-throated diver at (i) 3 cm resolution, (ii) 2 cm resolution and (iii) 1.5 cm resolution

Due to its size, and requirements to provide baseline information across a large area, the OPA is surveyed using a transect design (Fig. 4). The nearshore section of the survey area is surveyed along transects parallel to the shoreline to accommodate air traffic restrictions, and the offshore part of the survey area is

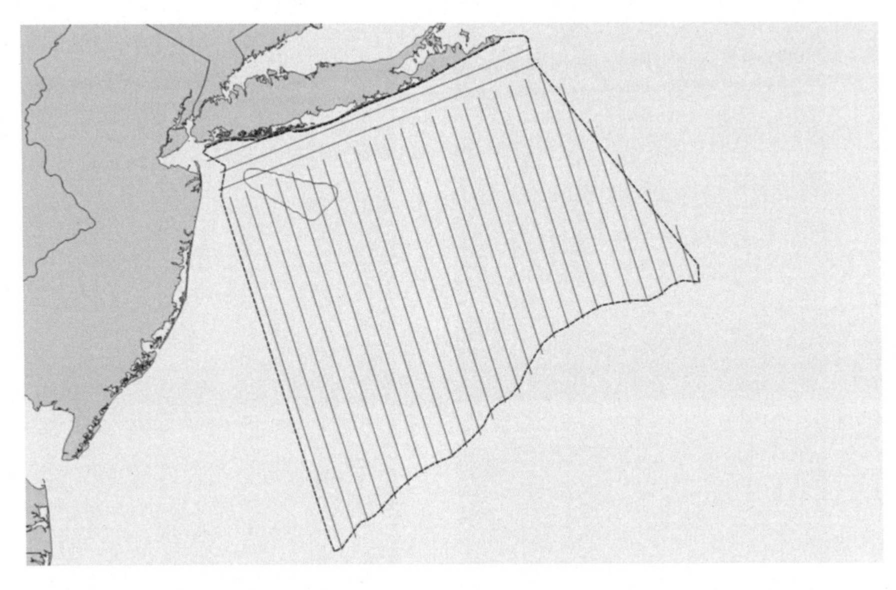

Fig. 4 Example transect lines for the offshore planning area (dashed line) and wind energy area (solid line) showing the orientation of the offshore planning area transect lines nearshore and offshore

surveyed along transects perpendicular to the shoreline and consequently to the bathymetry, providing optimal orientation for expected clines in the distribution of target species. The WEA surveys (required to provide more detailed information as part of the BACI survey) are conducted using a grid design that provides even coverage across the WEA and its associated buffer.

The resulting georeferenced targets are used to assess the distribution and abundance of birds, marine mammals, turtles and fish. The data collected in the surveys of the OPA and WEA are regularly uploaded to the open-access internet domain ReMOTe (https://remote.normandeau.com/nys_overview.php). The analysts and taxon-specific experts log into ReMOTe to access the raw data, identify species and interpret results. As data review progresses, survey results are publicly available on ReMOTe in near-real time. This comprehensive baseline study using an innovative survey technique will facilitate a more efficient planning of energy production offshore by providing the necessary information to meet regulatory requirements for environmental review of WEAs prior to the areas being designated as potential offshore wind areas.

3 Conclusion

The NYSERDA digital aerial surveys are setting new standards for offshore surveying, utilising a purpose-built camera system to produce ultrahigh-resolution images. This innovative camera system is helping to reduce consenting risk by allowing similar species with different consenting risks to be identified more readily. The trade-off between resolution and coverage often means higher resolutions are too costly to implement on a large scale. However, developments in technologies now provide for greater swath width and higher resolutions, meaning the cost per unit flying time is reduced. Utilising these innovations to undertake surveys on such a large scale such as the NYSERDA surveys is paving the way for the future standards of offshore surveying.

References

1. European Commission: Communication from the Commission to the European Parliament, The Council, The European Economic and Social Committee, The Committee of the Regions and the European Investment A Framework Strategy for a Resilient Energy Union with a Forward-Looking Climate Change Policy/* COM/2015/080 final */. (2014)
2. PD 6900:2015: Environmental impact assessment for offshore renewable energy projects – Guide. Innovate UK, Swindon (2015)
3. Camphuysen, K., Fox, A., Leopold, M., Petersen, I.: Towards standardised seabirds at sea census techniques in connection with environmental impact assessments for offshore wind farms in the U.K.: a comparison of ship and aerial sampling methods for marine birds, and their applicability to offshore wind farm assessments, NIOZ report to COWRIE (BAM – 02-2002), Texel (2004)
4. Rehfisch, M., Michel, S.: Workshop: techniques for surveying MPAs from the air – from bay to ocean. IMPAC 3 Proceedings, pp. 39-54, Marseille-Ajaccio (2013)
5. Coppack, T., McGovern, S., Rehfisch, M., Clough, S.: Estimating wintering populations of waterbirds by aerial high-resolution imaging. Vogelvelt. **137**, 149–155 (2017)
6. Thaxter, C., Burton, N.: High Definition Imagery for Surveying Seabirds and Marine Mammals: A Review of Recent Trials and Development of Protocols. British Trust for Ornithology Report Commissioned by Cowrie Ltd (2009)
7. Normandeau Associates, Inc.: High-resolution Aerial Imaging Surveys of Marine Birds, Mammals, and Turtles on the US Atlantic Outer Continental Shelf—Utility Assessment, Methodology Recommendations, and Implementation Tools. US Dept. of the Interior, Bureau of Ocean Energy Management. Contract # M10PC00099 (2012)
8. Ross, K., Burton, N., Balmer, D., Humphreys, E., Austin, G., Goddard, B., Schindler-Dite, H., Rehfisch, M.: Urban Breeding Gull Surveys: A Review of Methods and Options for Survey Design BTO Research Report 680. BTO, Thetford (2016)
9. Bakó, G., Tolnai, M., Takács, Á.: Introduction and testing of a monitoring and colony-mapping method for waterbird populations that uses high-speed and ultra-detailed aerial remote sensing. Sensors. **14**, 12828–12846 (2014)
10. Mellor, M., Maher, M.: Full Scale Trial of High Definition Video Survey for Offshore Windfarm Sites. COWRIE Ltd, London (2008)
11. Buckland, S., Burt, M., Rexstad, E., Mellor, M., Williams, A., Woodward, R.: Aerial surveys of seabirds: the advent of digital methods. J. Appl. Ecol. **49**, 960–967 (2012)
12. State of the State.: https://www.nyserda.ny.gov/About/Newsroom/2018-Announcements/2018-01-02-Governor-Cuomo-Unveils-20th-Proposal-of-2018-State-of-the-State (2018)

Wind Farm Effects on Migratory Flight of Swans and Foraging Distribution at Their Stopover Site

Sachiko Moriguchi, Haruka Mukai, Ryosuke Komachi, and Tsuneo Sekijima

Abstract Wind farms have unintended negative consequences for birds, such as bird collisions, habitat loss, and barrier effects. Japanese law now requires environmental impact assessments (EIAs) of wind farm construction. Despite these EIAs, assessments of wind farm effects on birds are often inadequate because no data are available that compare bird behavior and distribution before and after wind farm development. Here we investigated macro avoidance and the foraging distribution of swans before and after the construction and onset of operations of a wind turbine operation in Japan's Tohoku region. During the spring and fall migratory seasons, we used fixed-point observations to survey swan flight trajectories near a newly constructed wind farm and an existing, operational wind farm. Swan turning radius and trajectory altitude were used to determine macro avoidance of wind farms. Swan foraging distribution around wind farms was surveyed by car. Sightings of migratory swans drastically decreased in the wind farm areas, but swan foraging distribution around the turbines remained unaffected. This outcome may be because the wind farm is distant enough from existing swan foraging areas. We conclude that collision risk should be low because migrating swans avoided wind turbines, but their traveling distance is increased by the need to fly around the wind farm area.

Keywords Macro avoidance · Migration · Wind turbine · *Cygnus*

S. Moriguchi (✉)
Faculty of Agriculture, Niigata University, Niigata, Japan

Current address:
Faculty of Veterinary Science, Nippon Veterinary and Life Science University, Tokyo, Japan
e-mail: smoriguchi@nvlu.ac.jp

H. Mukai · R. Komachi
Graduate School of Science and Technology, Niigata University, Niigata, Japan
e-mail: f15n009c@mail.cc.niigata-u.ac.jp

T. Sekijima
Faculty of Agriculture, Niigata University, Niigata, Japan
e-mail: sekijima@gs.niigata-u.ac.jp

R. Bispo et al. (eds.), *Wind Energy and Wildlife Impacts*,
https://doi.org/10.1007/978-3-030-05520-2_8

1 Introduction

Although wind power is frequently promoted as a promising source of renewable energy, wind farms have had unintended negative effects on animals. The most direct consequence is collision with turbine blades, reported in many bird and bat species [1–6]. Indirect consequences include habitat loss [7–9] and related alterations in migration routes (i.e., barrier effects) [10].

In response, the Japanese government now requires wind farms that produce over 10,000 kW to perform environmental impact assessments (EIAs). Prior to the construction of any wind farm, power producers have to assess factors such as migration routes, species distribution, and collision risk for flying vertebrates [11, 12]. However, survey implementation is problematic because, while power producers plan fixed-point observations and line censuses based on recommendations in government-published manuals [11, 12], individual companies decide on details such as survey length, date, and time of day. As a result, EIA reliability is inconsistent across wind farms. Furthermore, post-construction surveys are unenforceable.

Very few data are available in Japan regarding the wider, indirect effects of wind farms, as most studies focus on quantifying bird collisions by examining carcasses [13, 14]. Habitat loss [15] and barrier effects are comparatively understudied. Importantly, research and EIAs have not compared changes in bird behavior and distribution before and after wind farm construction/operation.

One bird taxon that seems likely to experience indirect wind farm effects is Anatidae (ducks, geese, and swans). Reports from several countries indicate that these birds face considerable pressure from barrier effects and loss of foraging habitat. Wind farms on existing migratory routes cause geese and ducks to alter their travel paths [10, 16, 17]. A similar effect was observed in Bewick's swans (*Cygnus columbianus bewickii*) on roosting routes [9]. The avoidance of wind turbines built on former foraging areas also effectively limited the amount of usable habitat by geese and resulted in foraging habitat loss [20]. These movements are considered macro-avoidance behaviors, defined as staying away from a broad area around wind farms, in contrast to micro behaviors (directly avoiding turbine rotor blades) [18, 19]. As can be seen from these categorizations, determining whether birds exhibit macro avoidance involves some ambiguity and subjectivity. Despite this complication, more studies should investigate macro avoidance in Anatidae, because it appears to be the primary response of these birds to wind farms, meaning they are more likely to be affected by barrier effects and habitat loss stemming from such behaviors.

The purpose of this study is to assess changes in swan flight behavior and foraging distribution before wind farm construction, as well as before and after the start of operations. We selected whooper swans (*Cygnus cygnus*) and tundra swans (*Cygnus columbianus jankowskii*) as subjects. Both are major swan species in Japan that have limited wintering and stopover sites, making them particularly sensitive to changes in migratory routes and decreasing habitat range.

2 Methods

2.1 Study Area

The study site included two wind farms in southern Akita prefecture, Japan (Fig. 1). The site is on a flyway to wintering and stopover sites along the Sea of Japan for both study species. The old wind farm (a row of 15 wind turbines of 2000 kW) has been active since 2004, while the new one (a row of 17 wind turbines of 3000 kW) began its operation in December 2015. Both wind farms were constructed along mountain ridges. Swans foraged in plains surrounded by these mountains during stopover and wintering seasons. The wind farm areas (all rotor regions and surroundings of the wind turbines) were 0.608 km^2 and 2.641 km^2 for the old and new wind farms, respectively.

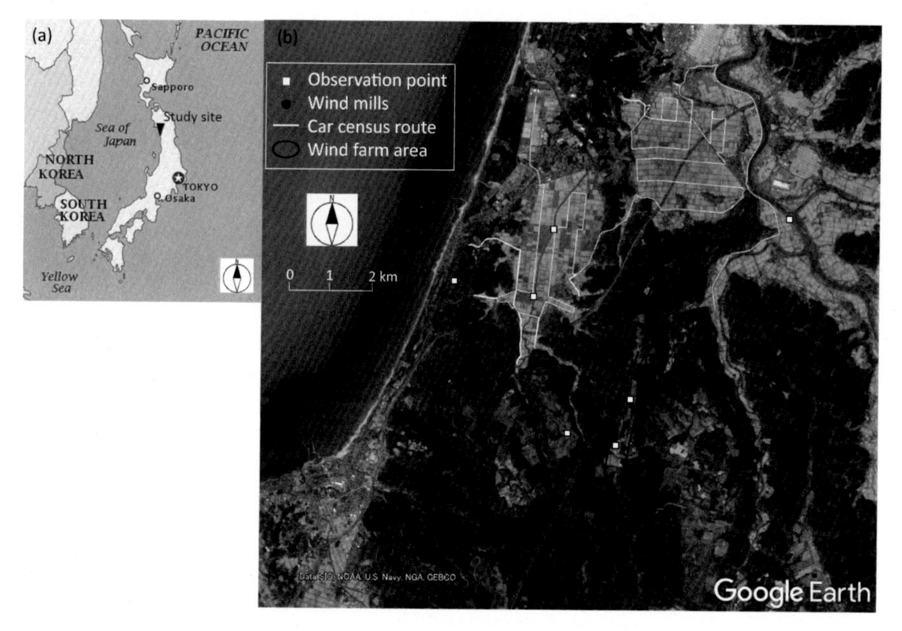

Fig. 1 Study area map. The black triangle shows the study site location (**a**). White squares are observation points, and black dots are wind turbines in the study area (**b**). White lines are the car census routes for observing the foraging distribution of swans. Areas outlined with black lines are wind farm areas. The eastern wind farm is an old wind farm with 15 wind turbines in operation since 2004, and the western one is a new wind farm with 17 wind turbines in operation since December 2015

2.2 Field Study

Migratory flights were recorded with fixed-point observations. Seven observation sites were established to monitor swan flocks at both wind farms. It was only possible to conduct surveys at three sites simultaneously; therefore, from these seven sites, we selected three sites from which we could monitor all the wind farm areas at the same time. We recorded flight trajectories, flock size, and flight altitude with binoculars and telescopes. Flight trajectories were recorded in reference to the topography, buildings, roads, and power lines. At the new wind farm, power producers conducted a pre-construction survey for 6 days in early November 2008 and early March 2009. Pre-operation surveys were conducted for 25 days in fall 2015, from October to November. Surveys during operation were conducted for 13 days in February (spring) and 34 days from October to November (fall) in 2016.

A census covering 50 km by car was conducted in the swans' foraging habitat (plain) near both wind farms to assess habitat loss. Foraging flock sizes and swan locations were recorded for 11 days pre-operation in fall, and then for 4 days in spring and 11 days in fall during operation, during the same periods as the fixed-point observations.

2.3 Determination of Macro Avoidance

Macro-avoidance rate was calculated using the populations exhibiting horizontal and vertical avoidance and non-avoiding populations, for both wind farms, using the following formula:

$$\text{Macro-avoidance rate} = 1 - \frac{a}{a+b} \tag{1}$$

a: size of population passing above the wind turbine rotor area
b: size of population avoiding the rotor area horizontally and vertically

Our original calculation employed all flight trajectories suspected to pass through the rotor area at any altitude (Fig. 2). After entering a buffer zone, vertical-avoidance flocks flew higher or lower than the rotor area, while horizontal-avoidance flocks moved sideways. Unique to each species, the buffer zone radius was defined as the sum of the rotor radius and turning radius.

The turning radius is an aeromechanics variable [21] calculated from flight speed, gravitational acceleration, and the tangent of the banking angle, as follows:

$$\text{Turning radius} = \frac{\text{flight speed}^2}{\text{gravitational acceleration} \times \tan\theta} \tag{2}$$

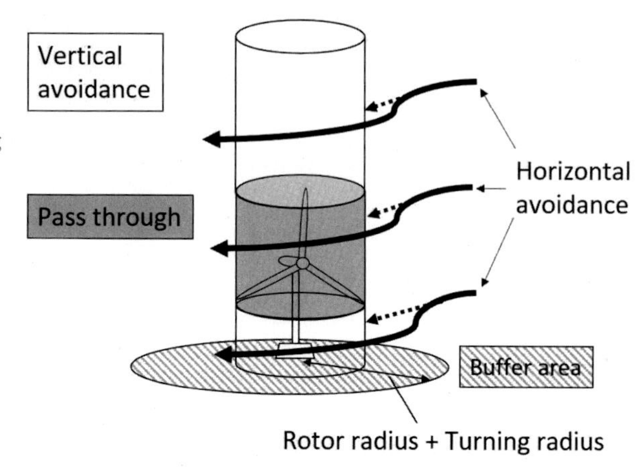

Fig. 2 Diagram of macro avoidance, divided into vertical and horizontal avoidance. The former comprises trajectories passing through the upper or lower rotor area (gray circular column), while the latter comprises trajectories (bold arrows) that shift in direction after entering the buffer area (striped region), defined as the sum of the rotor radius and the swan turning radius

Flight speed was 20 m/s [22, 23] and 14 m/s [23] for tundra and whooper swans, respectively. The faster value was selected as flight speed for use in the formula. A banking angle of 5° was determined through field observation, resulting in a turning radius of 467 m and a buffer zone radius of 517 m.

2.4 *Statistical Analysis*

Welch's two-sample *t*-test with Bonferroni corrections was used to examine differences in flight-flock size within and outside the two wind farm areas. Differences in the proportions of avoidance and non-avoidance in swan populations between pre-construction and before and during operation of the new wind farm were verified using Fisher's exact test, followed by Bonferroni corrections. The distances of the foraging swan flocks from the new wind farm before and during operation were analyzed with a Wilcoxon rank sum test. The latter was also used to examine between season differences in foraging flock distance from the new wind farm during operation.

All spatial data were analyzed in ArcGIS 10.4.1 [24], and all statistics were conducted in R 3.2.3 [25].

3 Results

Throughout the entire study site, we recorded 146, 121, and 628 flight populations of swans in fall 2015, spring 2016, and fall 2016, respectively. Within the wind farm area, we observed 24, 28, 10, 5, and 30 swan flocks in fall 2008, spring 2009, fall 2015, spring 2016, and fall 2016, respectively. It was not possible to identify the swan species on migratory flights. After the new wind farm was constructed,

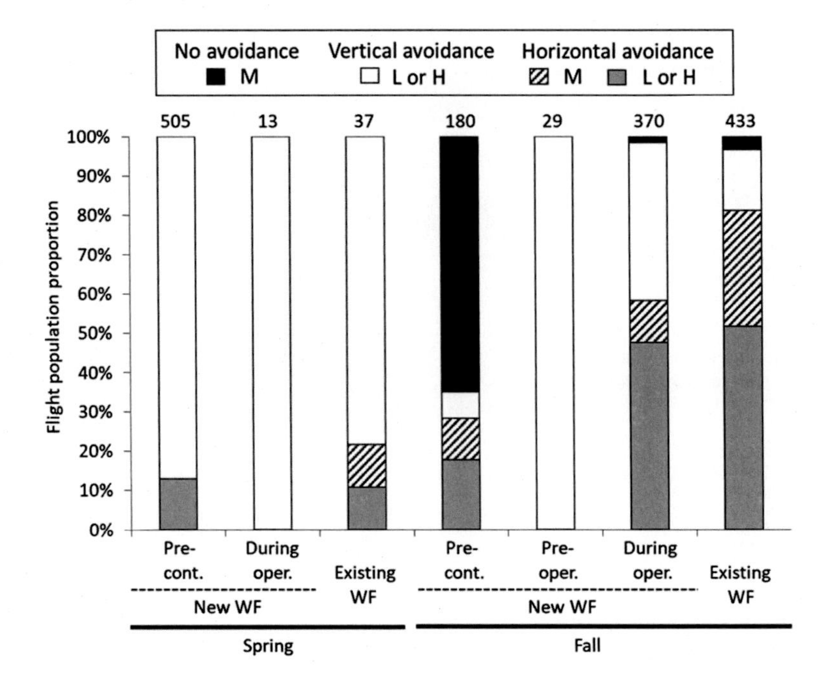

Fig. 3 Proportion of flight population engaging in macro avoidance versus no avoidance per season and wind farm (WF). Pre-cont., pre-construction flight population; pre-oper, pre-operation; during oper., during operation of the new WF. Black bars indicate non-avoidance flight at the rotor altitude (M). White bars indicate non-avoidance flight at lower or higher altitudes (L or H). Striped bars indicate horizontal avoidance at M, and gray bars indicate avoidance at lower or higher altitudes. Numbers above the bars indicate flight-flock sample size

the flock size of flying swans within the rotor area was 5.13 ± 2.42 (SD), which was significantly smaller than that outside the rotor area (16.84 ± 10.29; $t = 6.62$, $df = 49$, $P < 0.001$) and smaller than those at rotor area height before wind farm construction (19.68 ± 20.11; $t = 3.19$, $df = 18.9$, $P = 0.005$) and outside the wind farm area (10.49 ± 14.75; $t = -5.23$, $df = 40.1$, $P < 0.001$).

In spring, most swans flew higher than in fall, resulting in a high avoidance proportion even before wind farm construction ($P < 0.001$, Fig. 3). In fall, the avoidance proportion was 35% pre-construction, 100% before operation ($P < 0.001$), and 98% during operation ($P < 0.001$). During operation flight behavior was similar to behavior at the old wind farm. The two swan species often foraged in mixed species flocks; therefore species-specific distances to the new wind farm were not considered.

Foraging distance from the new wind farm did not differ between before and during operation in fall ($W = 3726$, $P = 0.89$). However, a seasonal difference was present in the foraging distance during operation ($W = 1611$, $P < 0.001$) (Fig. 4).

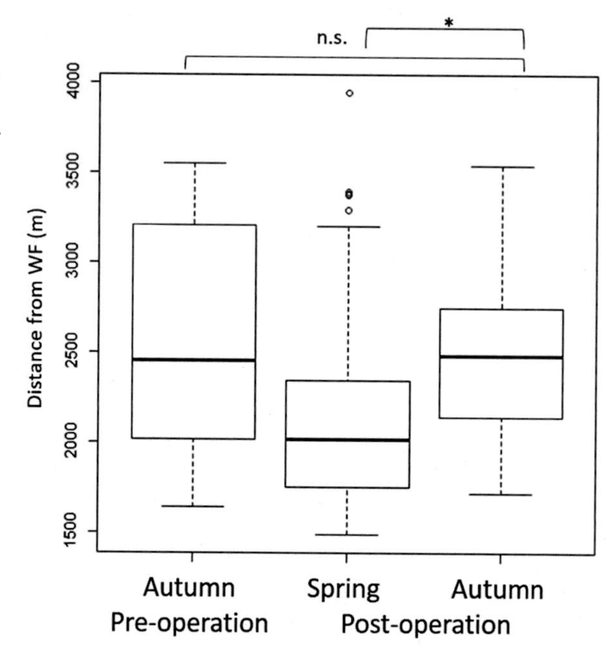

Fig. 4 Distance of foraging swan flocks from the nearest wind turbine at the new wind farm (WF), separated by season and operational status. *refers to a significant difference and *n.s.* indicates "not significant"

4 Discussion

This study successfully demonstrated how wind farms can affect migrating swans. First, the flock size of migratory swans decreased after wind farm construction. Our results corroborate previous studies that indicated a decrease in swan and flock numbers in wind farm areas, although they did not determine the exact flock size [5, 9]. Thus, our study is the first to confirm that swan population sizes in wind farm areas decrease because of both smaller flock sizes and a lower overall number of flocks.

Most swans avoided wind turbines before entering the wind farm area, in line with previous reports of a low collision risk for these birds [5, 13]. In addition, we observed seasonal differences in swan flight altitude, probably due to terrain-related factors. In spring, swans flew over the mountains before reaching foraging grounds near the wind farm, which required higher altitudes. In contrast, during fall, swans flew over the plain and could maintain lower altitudes.

Swan foraging flocks did not avoid the new wind farm before or during operation. The foraging area was approximately 2 km away from the wind farm, suggesting that foraging swans were far enough from the farm to be buffered from any negative effects. Indeed, previous research indicated that swans and geese can safely forage from several thousand meters to 20 m away from wind turbines [5, 20, 26, 27]. Furthermore, the location of the wind farms on a mountain ridge may have contributed to their lack of influence, as swans forage on the plains between the mountains. Thus, future studies should assess the environmental impact of wind farms on plains [e.g., 20, 27], as turbines directly on foraging fields would likely result in severe habitat loss.

Despite having very little pre-construction data due to the limitations in the power company's assessment, we nonetheless observed clear differences in pre- and post-construction migratory flights. In conclusion, our study revealed that swans are at low risk of collision with wind turbines but are still subjected to barrier effects. We therefore recommend further investigation on energy loss in swans that must alter their travel to fly around wind farms.

References

1. Barclay, R.M.R., Baerwald, E.F., Gruver, J.C.: Variation in bat and bird fatalities at wind energy facilities: assessing the effects of rotor size and tower height. Can. J. Zool. **85**, 381–387 (2007). https://doi.org/10.1139/Z07-011
2. Smallwood, K.S., Thelander, C.: Bird mortality in the Altamont Pass Wind Resource Area, California. J. Wildl. Manag. **72**, 215–223 (2008). https://doi.org/10.2193/2007-032
3. Drewitt, A.L., Langston, R.H.W.: Collision effects of wind-power generators and other obstacles on birds. Ann. N. Y. Acad. Sci. **1134**, 233–266 (2008). https://doi.org/10.1196/annals.1439.015
4. Newton, I., Little, B.: Assessment of wind-farm and other bird casualties from carcasses found on a Northumbrian beach over an 11-year period. Bird Study. **56**, 158–167 (2009). https://doi.org/10.1080/00063650902787767
5. Rees, E.C.: Impacts of wind farms on swans and geese: a review. Wild. **62**, 37–72 (2012)
6. Amorim, F., Rebelo, H., Rodrigues, L.: Factors influencing bat activity and mortality at a wind farm in the Mediterranean region. Acta Chiropterologica. **14**, 439–457 (2012). https://doi.org/10.3161/150811012X661756
7. Pearce-Higgins, J.W., Stephen, L., Langston, R.H., Bainbridge, I.P., Bullman, R.: The distribution of breeding birds around upland wind farms. J. Appl. Ecol. **46**, 1323–1331 (2009). https://doi.org/10.1111/j.1365-2664.2009.01715.x
8. Garvin, J.C., Jennelle, C.S., Drake, D., Grodsky, S.M.: Response of raptors to a windfarm. J. Appl. Ecol. **48**, 199–209 (2011). https://doi.org/10.1111/j.1365-2664.2010.01912.x
9. Fijn, R.C., Krijgsveld, K.L., Tijsen, W., Prinsen, H.A.M., Dirksen, S.: Habitat use, disturbance and collision risks for Bewick's Swans *Cygnus columbianus bewickii* wintering near a wind farm in the Netherlands. Wildfowl. **62**, 97–116 (2012)
10. Masden, E.A., Haydon, D.T., Fox, A.D., Furness, R.W., Bullman, R., Desholm, M.: Barriers to movement: impacts of wind farms on migrating birds. ICES J. Mar. Sci. **66**, 746–753 (2009). https://doi.org/10.1093/icesjms/fsp031
11. Ministry of the Environment, Japan: Environmental Assessment Manual for Improvement of Wind Farm Site Location About Birds and Other Animals. Ministry of the Environment, Japan (2011). (in Japanese)
12. Ministry of Economy, Trade and Industry, Japan: Environmental Assessment Manual for Power Plants. Ministry of Economy, Trade and Industry, Tokyo (2017). (in Japanese)
13. Ura, T.: Cases of wind turbine impacts on birds in Japan. Strix. **31**, 3–30 (2015). (in Japanese with English abstract)
14. Kitano, M., Shiraki, S.: Estimation of bird fatalities at wind farms with complex topography and vegetation in Hokkaido, Japan. Wildl. Soc. Bull. **37**, 41–48 (2013). https://doi.org/10.1002/wsb.255
15. Takeda, K.: The impact of wind turbines on breeding bird densities. Japan. J Ornithol. **62**, 135–142 (2013). https://doi.org/10.3838/jjo.62.135 (in Japanese with English abstract)

16. Desholm, M., Kahlert, J.: Avian collision risk at an offshore wind farm. Biol. Lett. **1**, 296–298 (2005). https://doi.org/10.1098/rsbl.2005.0336
17. Plonczkier, P., Simms, I.C.: Radar monitoring of migrating pink-footed geese: behavioural responses to offshore wind farm development. J. Appl. Ecol. **49**, 1187–1194 (2012). https://doi.org/10.1111/j.1365-2664.2012.02181.x
18. Cook, A.S.C.P., Humphreys, E.M., Masden, E.A., Burton, N.H.K.: The avoidance rates of collision between birds and offshore turbines. Scott. Mar. Freshwat. Sci. **5**(16) (2014). https://doi.org/10.1111/1365-2664.12191
19. Whitfield, D.P., Urquhart, B.: Deriving an avoidance rate for swans suitable for onshore wind farm collision risk modelling. Nat. Res. Inf. Note, 6 (2015)
20. Madsen, J., Boertmann, D.: Animal behavioral adaptation to changing landscapes: spring-staging geese habituate to wind farms. Landsc. Ecol. **23**, 1007–1011 (2008). https://doi.org/10.1007/s10980-008-9269-9
21. Ochiai, K.: Aeromechanics. Nihon Kouku Gijutsu Kyoukai, Tokyo (1999). (in Japanese)
22. Klaassen, M., Beekman, J., Kontiokorpi, J., Mulder, R.W., Nolet, B.: Migrating swans profit from favourable changes in wind conditions at low altitude. J. Ornithol. **145**, 142–151 (2004). https://doi.org/10.1007/s10336-004-0025-x
23. Yui, M., Shimada, Y.: A new sphere shape model for estimating the number of bird-wind turbine collisions. J Policy Stud. **15**, 1–17 (2013). (in Japanese with English abstract)
24. ESRI Inc.: ArcGIS 10.4.1, Redlands, CA (2015)
25. R Development Core Team. R: A Language and Environment for Statistical Computing. R Foundation for Statistical Computing, Vienna (2015)
26. Larsen, J.K., Madsen, J.: Effects of wind turbines and other physical elements on field utilization by pink-footed geese (*Anser brachyrhynchus*): a landscape perspective. Landsc. Ecol. **15**, 755–764 (2000). https://doi.org/10.1023/A:1008127702944
27. Harrison, A., Hilton, G.: Fine-Scale Distribution of Geese in Relation to Key Landscape Elements in Coastal Dobrudzha, Bulgaria Preliminary report. WWT, Slimbridge (2014)

Automatic Bird Identification for Offshore Wind Farms

Juha Niemi and Juha T. Tanttu

Abstract There is a need for automatic bird identification system at offshore wind farms in Finland. The developed system should be able to operate from onshore, which is cost-effective in terms of installations and maintenance. Indubitably, a radar is the obvious choice to detect flying birds, but external information is required for actual identification. A conceivable method is to exploit visual camera images. In the proposed system, the radar detects birds and provides the coordinates to camera steering system. The camera steering system tracks the flying birds, thus enabling capturing a series of images. Classification is based on the images, and it is implemented by a small convolutional neural network trained with a deep learning algorithm. We also propose a data augmentation method in which images are rotated and converted in accordance with the desired color temperatures. The final identification is based on a fusion of data provided by the radar and image data. We present the results of the number of correctly identified species based on manually taken images.

Keywords Image classification · Deep learning · Convolutional neural networks · Machine learning · Data augmentation

1 Introduction

The first offshore wind farm is under construction on the Finnish west coast. The minimal demand of the environmental license defines that bird species in the turbine areas are monitored and possible collisions prevented concerning especially two species: the white-tailed eagle (*Haliaeetus albicilla*), and the lesser black-backed

J. Niemi (✉)
Signal Processing Laboratory, Tampere University of Technology, Pori, Finland
e-mail: juha.k.niemi@tuni.fi

J. T. Tanttu
Mathematics Laboratory, Tampere University of Technology, Pori, Finland
e-mail: juha.tanttu@tuni.fi

R. Bispo et al. (eds.), *Wind Energy and Wildlife Impacts*,
https://doi.org/10.1007/978-3-030-05520-2_9

gull (*Larus fuscus fuscus*). At present, the radar system controls directly the pilot wind turbine and shut it down when any bird flies into a perimeter of 300 m to the wind turbine, which is the minimum distance in terms to have sufficient time to shut down the turbine. The number of fast restarting of the turbines should stay as low as possible due to wearing of the mechanics, and therefore this operation should be used as a last resort. A solution to this is a suitable deterrent method, but it is difficult to find only one applicable deterrence for all bird species in the wind farm area, and one extra problem is that the breeding birds may quickly become accustomed to, e.g., sounds as a deterrent method [1]. At first stage, an automatically operating bird species identification system is needed in order to be able to develop such deterrent system. The final objective is to develop a deterrent system that operates on species-group level (i.e., a different deterrent method for, e.g., gulls and eagles). At this point our main research question is how to identify bird species in flight automatically in real time.

It makes sense, in order to implement this system cost-effectively, to build only one control system in such a location from where it is possible to monitor birds in the vicinity of all wind turbines of the wind farm. To achieve this goal, it has been decided to place the control system onshore as it is more cost-effective compared to a system installed offshore. The distance from the chosen location to the monitored birds (i.e., the vicinities of the future wind turbines) ranges between 500 and 1500 m. We initially considered a radar for both detecting the bird and identifying its species. However, it is known that the identification capacity is limited, rendering impossible to classify bird species any further merely by this radar system. Obviously, external information is required, and a conceivable method is to exploit visual camera images. A digital single-lens reflex (DSLR) camera with telephoto lens is applied due to the long photographing distance.

2 The System

The proposed system consists of several hardware as well as software modules. See Fig. 1 for an illustration. First of all there is the radar which is connected to a local area network (LAN), and thus it is able to communicate with the servers in which the various programs are running. The most important role of the radar is to detect flying birds, but it also provides some parameters for bird identification (i.e., classification). The parameters are the distance in 3D of a target (m), velocity of a target (m/s), and trajectory of a target (WGS84 coordinates). The distance of a target is used to estimate the size in meters. Velocity of a target is used for the final classification, which we call the fusion. The system also includes the aforementioned camera with the lens and a motorized video head. The video head is operated by Pelco-D control protocol [2], and the control software for it is developed by us. The camera is controlled by the application programmable interface (API) of the camera manufacturer, and the software for controlling the camera is also developed by us. The system has three servers: the radar server, the video head

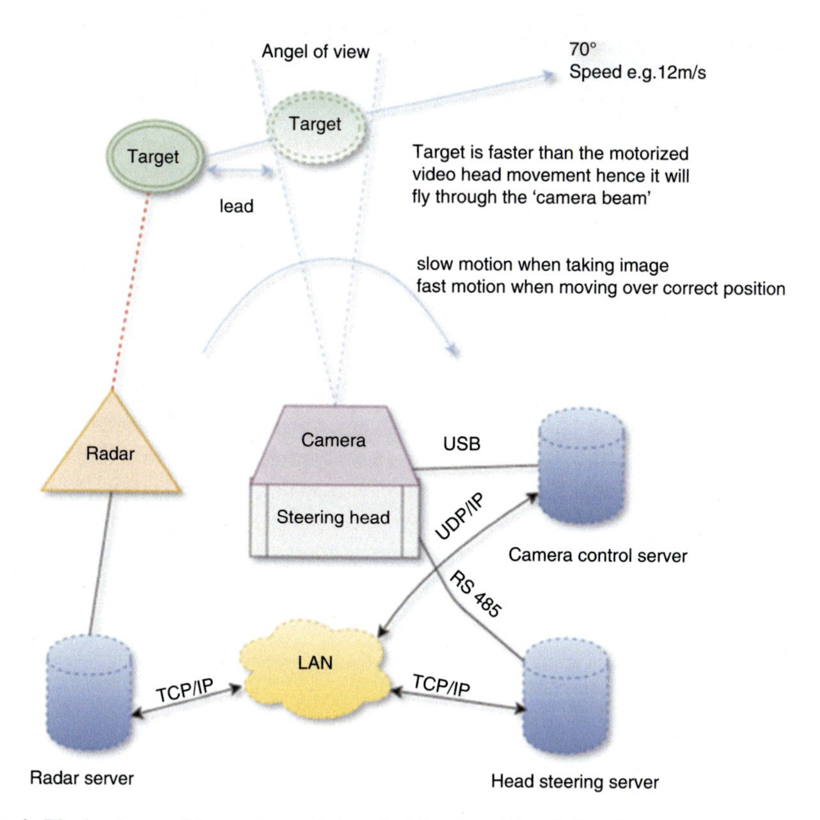

Fig. 1 The hardware of the system and the principle of catching flying bird into the frame area of the camera

steering server, and the camera control server. Software for the radar server is supplied by the manufacturer of the radar, but the software for the other two servers is result of our development work.

2.1 Hardware Level

Figure 1 illustrates the hardware components of the system. The operation on hardware level commences when the radar detects a bird and sends information of the blip including the parameters to the video head steering server. The video head steering server reads the coordinates from the data sent by the radar. The system reacts only if the data has the altitude information of the object.

We have used the PT-1020 Medium Duty video head of the 2B Security Systems. This head has five different speeds of panning, 12, 18, 24, 32, and 48, as well as 2 of tilting, 12 and 18, in degrees/s. The maximum speed is 1499 m/s (i.e., 48 degree/s) at the distance of 1350 m, where the pilot turbine is located, and the minimum speed is 287 in m/s (i.e., 12 in degree/s) . The maximum ground speed of a flying bird in the test area is estimated to be 110 km/h = 30.5 m/s [3] added with the maximum average wind of 30 m/s [4] resulting in 60.5 m/s. The minimum ground speed of a flying bird in the test area is roughly 6 m/s, which is based on the radar. Soaring gulls and terns are not included in the minimum speed measure. The calculations show that the maximum speed of the head is sufficient, but the minimum speed is too fast for direct tracking.

The too-fast minimum speed problem (i.e., the minimum speed of steering the video head is not slow enough in order to track a flying bird at a desired distance) is solved as follows: the video head steering system calculates the most probable trajectory of the target bird and adds a lead to the calculated position so that when the head will be driven to that final position, it would be ahead of the target bird. The calculations are based on the flight speed (measured by the radar) of a target and the right-angled triangle, and they ignore any possible sudden deviations off the flight path. The curvature of the horizon is also ignored due to the relatively short distances in question. When the final position is reached, the head stops. This solves the minimum speed problem and also maximizes the probability that the target bird is within the frame of the camera at distance in question; see Fig. 1. After the head is driven to the final position, the camera takes series of images and sends them to the classification software that runs on the same server.

2.1.1 Radar

We use a radar system supplied by Robin Radar Systems B.V. because they provide an avian radar system that is able to detect birds. They also have tracker algorithms for tracking a detected object over time, i.e., between the blips. The model we use is the ROBIN 3D FLEX v1.6.3, and it is actually a combination of two radars and a software package for implementation of various algorithms such as the tracker algorithms. The two radars are a horizontal scanning S-band radar and a 3D tracking frequency-modulated continuous wave (FMCW) vertical X-band radar. The 3D tracking FMCW radar supports two-axis scanning mode for 3D coverage and tracking mode with either manual or automatic track selection. The ROBIN 3D FLEX v1.6.3 radar system is capable to provide parameters, such as velocity and bearing of the detected object, for our system. The S-band radar enables a long-range detection up to 10 km of flying birds, and it provides the longitude and the latitude of a target bird. The X-band radar enables higher resolution, i.e., it can detect smaller objects such as small birds up to distance of 5 km. The X-band radar also provides the altitude of the target [5–7].

The S-band radar can operate the whole 360°, but it is adjusted to operate 180° in the test site since the objects are always in the sector of south via west to north. The X-band radar operates roughly in 20° sector that can be configured to lie in a constant position, or it can be configured to multiplex between two separate positions. The two possible operation modes of the X-band radar are presented in Figs. 2 and 3, respectively. The ability to multiplex is important because it enables to adjust the minimum distance to the wind turbine of the approaching bird, and

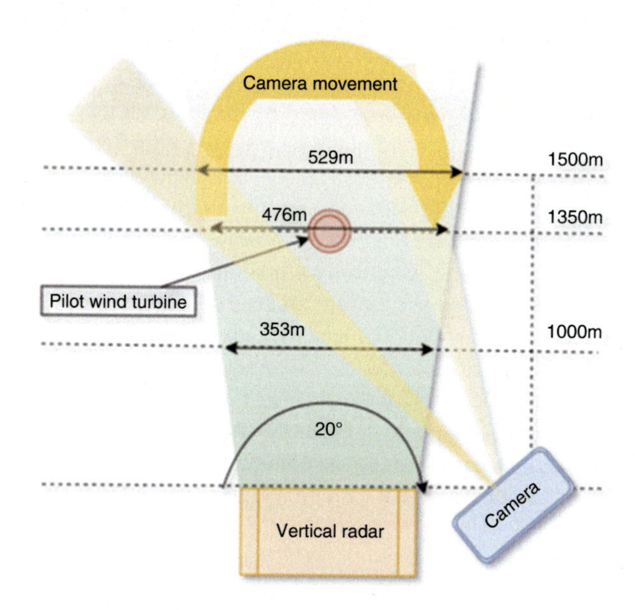

Fig. 2 The vertical X-band radar operation at a constant position

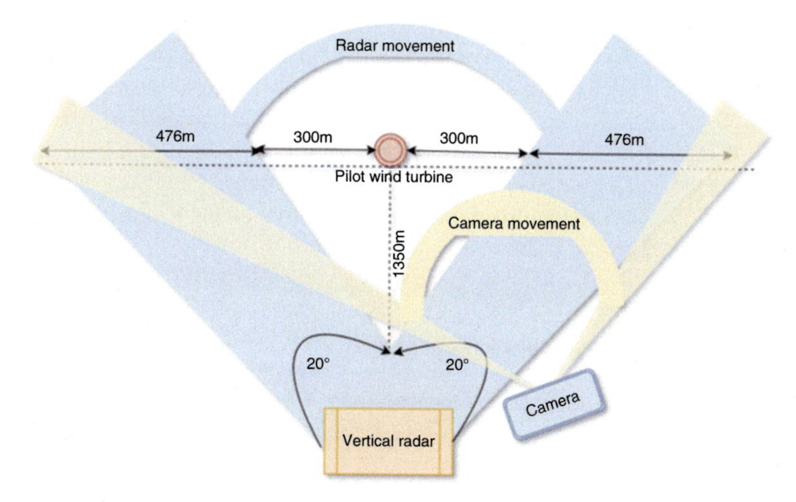

Fig. 3 The vertical X-band radar operation when configured to multiplex between two separate positions

therefore it gives sufficient time to shut down the wind turbine. The cross section of the two radar beams is the key feature since the intersection of this is the area from where all the three coordinates (i.e., latitude, longitude, and altitude) are available [5–7]. The proposed system cannot work without information about the altitude of the object.

2.1.2 Camera System

The resolution of the camera sensor measured by the total number of pixels and the focal length of the lens are important qualities because of the long distance to birds of which images are to be taken. We use a term called the effective number of pixels (ENP) defined by the number of pixels representing a bird. The remaining number of pixels is considered noise. Image classification is achieved only with ENP as birds will be very small (i.e., they consist of only a small number of pixels) in the images. ENP depends on the focal length of the lens and can be increased by choosing a long (i.e., in terms of focal length) telephoto lens. Because of these facts, we have chosen to use the Canon EOS 7D II camera with 20.2-megapixel sensor and the Canon EF 500/f4 IS lens. Correct focusing of the images relies on the autofocus system of the lens and the camera. Automatic exposure is also applied. The operation of the proposed system is not restricted to this combination of the camera and the lens, but a combination of any standard DSLR camera with any standard lens suitable for that camera can be utilized.

2.2 *Software Level*

The ability of the system to identify bird species depends on the image classification and the size and velocity estimates of the target bird. Figure 4 illustrates the work flow for processing and classifying images. The classification process begins with a series of images taken of a single bird (i.e., a bird individual or a tight flock of birds flying past). The purpose of taking several images is simply to increase the probability of a correct identification. The images in a single series are fed to the classifier separately.

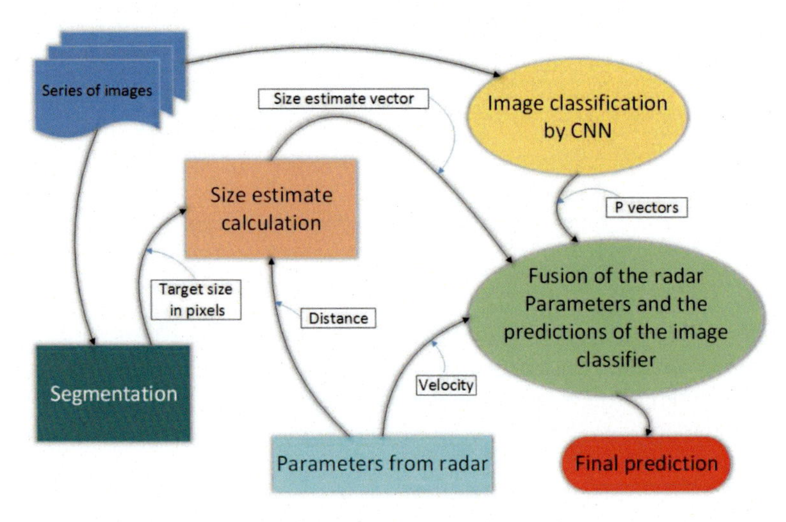

Fig. 4 The work flow for processing and classifying images

Segmentation is computed in parallel to image classification in order to obtain an estimate of the target bird size in pixels. We tested several methods from simple threshold to fuzzy logic for solving the problem at hand, i.e., a dark figure against bright background and *vice versa*. The background and the target can share several colors in the RGB color space. We achieved the best results, in terms of signal-to-noise ratio, by applying fuzzy logic segmentation [8]. In particular, we applied Mamdani's fuzzy inference method [9].

Once the target bird size in pixels is obtained, the target bird size in meters can be estimated. The image classifier results are given in $\boldsymbol{P}$-vectors, which elements are probabilities of belonging to each class. There is one $\boldsymbol{P}$-vector for each image. The $\boldsymbol{P}$-vectors for a single series of images are combined to the final $\boldsymbol{P}$-vector by adding element-wisely. A size estimate and velocity of a target bird are also represented as vectors. Finally, the fusion is simply an element-wise multiplication of the size vector, the velocity vector, and the combined $\boldsymbol{P}$-vector.

2.3 Data

Input data for the identification system consist of digital images and parameters from the radar. The parameters from the radar are real numbers such as velocity of a flying bird in m/s and bearing (i.e., a heading: the horizontal angle between the direction of an object and that of true north) in degrees.

Deep learning is a subfield of machine learning concerned with algorithms inspired by the structure and function of the brain called artificial neural networks. Deep learning networks are typically large, i.e., they have many layers and large number of parameters. A convolutional neural network (CNN) is one implementation of deep learning [10]. A CNN is a specialized kind of neural network for processing data that has a known grid-like topology like image data, which can be thought of as a 2D grid of pixels. Convolutional networks are simply neural networks that use convolution in place of general matrix multiplication in at least one of their layers. Typically CNNs need large data sets for the training phase to achieve sufficient classification performance [10].

All images for training the CNN have been taken manually at the test location and in various weather conditions. The location is the same where the camera will be installed for taking images automatically. There are also constraints concerning the area where the images have to be taken. Here the area refers to the air space in the vicinity of the pilot wind turbine. We have used the swept area (i.e., 130 m) as a suitable altitude level constraint for taking the images. At this stage, the images are only need to be taken in the vicinity of 1350 m in lengthwise direction, which is the distance to the pilot wind turbine. No normalization is applied to the images even though the sky at the background can be concolorous or multicolored.

The number of training examples for each image class (i.e., bird species) varies from 1164 to 5614. This is not enough (i.e., the demand is typically from tens of thousands to hundreds of thousands) for training of the deep learning algorithm [10], and therefore data augmentation (a.k.a. data expansion) is required. Data augmentation refers to any method that makes the original data set larger. These methods may include noise addition or various transformations such as rotation of images [11]. The number of images of each class should be the same as a CNN is applied and therefore the lowest number of images of the classes is used. We have restricted the number of classes (which includes both key species) to six due to difficulties in achieving large numbers of images for all possible classes in a relatively short period of time. The six initial classes for training the CNN are the common goldeneye (*Bucephala clangula*), the white-tailed eagle (*Haliaeetus albicilla*), the herring gull (*Larus argentatus*), the lesser black-backed gull (*Larus fuscus*), the great cormorant (*Phalacrocorax carbo*), and common/Arctic tern (*Sterna hirundo/paradisaea*). These species are chosen simply because we were able to collect the largest numbers of images of them. We are able to increase the number of classes (i.e., species) as the number of images is increasing in those classes (i.e., other than the aforementioned six) that have only minor number of images at present. The final classes will also include the common scoter (*Melanitta nigra*), the velvet scoter (*Melanitta fusca*), the common eider (*Somateria mollissima*), the common gull (*Larus canus*), and the black-headed gull (*Larus ridibundus*).

2.3.1 Data Augmentation

Our system is operating in marine environment, and therefore prevailing weather has significant influence to the tonality of the images taken at the test site. It is intuitively obvious that the lighting will vary with time, and thus the toning of the images will be changing according to lighting.

Color (in K, Kelvin degrees) temperature is a property of a light source. It is the temperature of the ideal blackbody radiator that radiates light of the same color as the corresponding light source. In this context blackbody radiation is the thermal electromagnetic radiation emitted by a black body. A blackbody is an opaque and nonreflective body. It has a specific spectrum and intensity that depend only on the temperature of the blackbody, and it is assumed to be uniform and constant. In our case the light source is the sun that closely approximates a blackbody radiator. Even though the color of the sun may appear different depending on its position, the changing of color is mainly due to the scattering of light, and it is not because of the changes in the blackbody radiation [12–15]. Color matching functions (CMFs) provide the absolute energy values of three primary colors which appear the same as each spectrum color. We applied the International Commission on Illumination (*Commission internationale de l'clairage*, CIE) 10-deg color matching functions in our data augmentation algorithm [16].

The augmentation is done according to the curves in Fig. 5. Mitchell Charity [17] by converting an image into different color temperatures between 2000 and 15,000 K with step size s, where $s \in \{25, 50, 75, 100, 150, 200, 250, 300, 1000\}$. This makes the training set significantly larger, e.g., if s is 200, the class with the smallest number of training examples of 1164 becomes $67 \cdot 1664 = 77{,}988$, where

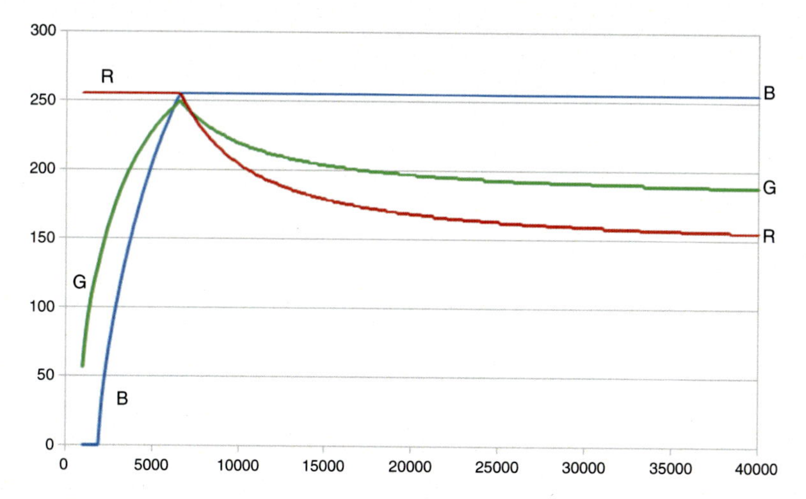

Fig. 5 Color temperature and corresponding RGB values presented according to CIE 1964 10-degree color matching function

67 = (15,000 – 2000)/200 + 2 (i.e., the difference between 15,000 and 2000 divided by the step size plus the one extreme). When conversion is done, the images are also rotated by a random angle between $-30°$ and $30°$ drawn from the uniform distribution. Motivation for this is that a CNN is invariant to small translations but not rotation of an image [18].

3 Classification

3.1 Target Size Estimate and Velocity

In this subsection we describe how the size of the target bird is estimated. The frame size ([width x height y], in *pixels*) of the camera and the angle of view (α) of the lens are known. The distance (d) to the target bird is provided by the radar. The maximum number of horizontal (σ_h) and vertical (σ_v) *pixels* of the target bird are calculated from the segmented image, respectively. The angle of view, b, at the distance, d, is calculated over a right-angled triangle, see Fig. 6. The horizontal number of *pixels/meter* is given by:

$$\rho_h = \frac{x}{b_h}, \tag{1}$$

and the vertical number of *pixels/meter* by:

$$\rho_v = \frac{y}{b_v}, \tag{2}$$

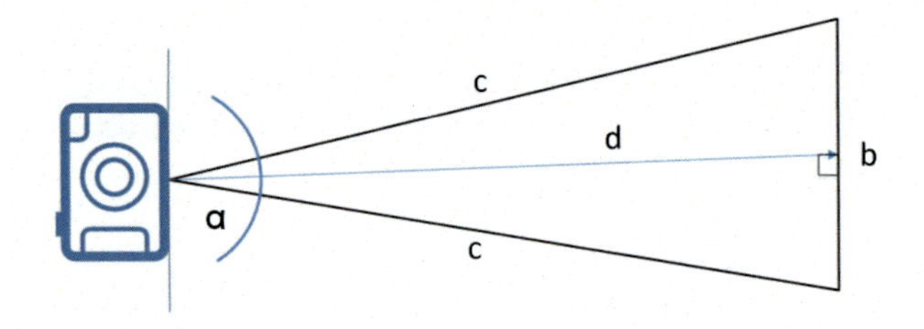

Fig. 6 Diagram of the size estimate calculation

where b_h and b_v denote the horizontal and the vertical angles of view, respectively. The estimate for the size of the bird in a single image in *square meters* as an area of rectangle is:

$$e = \frac{\sigma_h}{\rho_h} * \frac{\sigma_v}{\rho_v}. \tag{3}$$

The size estimate is presented as a vector with elements placed according to the class order (ordered by the probabilities), i.e., *class 1, class 2, class nc*, where *nc* denotes the number of classes. The composition of the vector is the following: calculate the average of the size estimates of the image series; check from the size look-up table for which classes the average size, *e*, fits; turn those elements to one; and set the others to zero, yielding:

$$\boldsymbol{E} = [e_1, e_2, \ldots, e_{\mathrm{nc}}], \tag{4}$$

with elements:

$$e_j = \begin{cases} 1, & \text{if } e \text{ fits class } j, \\ 0, & \text{otherwise.} \end{cases} \tag{5}$$

The velocity of the target bird is a parameter provided by the radar also presented as a vector. It is composed in similar way than the $\boldsymbol{E}$-vector in (4), i.e., check from the velocity look-up table for which classes the provided velocity, *v*, fits and turn those elements to one and the others to zero.

$$\boldsymbol{V} = [v_1, v_2, \ldots, v_{\mathrm{nc}}], \tag{6}$$

with elements:

$$v_j = \begin{cases} 1, & \text{if } v \text{ fits class } j, \\ 0, & \text{otherwise.} \end{cases} \tag{7}$$

3.2 Image Classification

Figure 7 illustrates image classification. We applied a CNN for feature extraction and built an architecture of three convolution layers of which each is followed by rectifier layers. The first two layers are followed by cross-channel normalization layers, and the second and third layer is followed by max-pooling layer. Finally, three consecutive fully connected layers are at the end of the CNN network. The dropout technique is applied to the first and second fully connected layers. This architecture may be seen as a small CNN because it is actually non-deep compared with conventional deep learning models [19].

We applied a two-step learning method, i.e., the CNN is trained with the first N-1 layers (i.e., N is the number of the layers) viewed as feature vectors, and a combined result is based on these feature vectors that is used to train a support vector machine (SVM) classifier [20]. The SVM classifier is thus the actual classifier [19].

The result of the image classifier is presented as a vector: thus predictions for each image in a single series of images are combined into the vector, $\boldsymbol{P}_i$:

$$\boldsymbol{P}_i = [c_1, c_2, \ldots, c_{\mathrm{nc}}], i = 1, \ldots, n, \tag{8}$$

where c_j is a probability of belonging to *class j*, *nc* is the number of classes, and n is the number of images in each series.

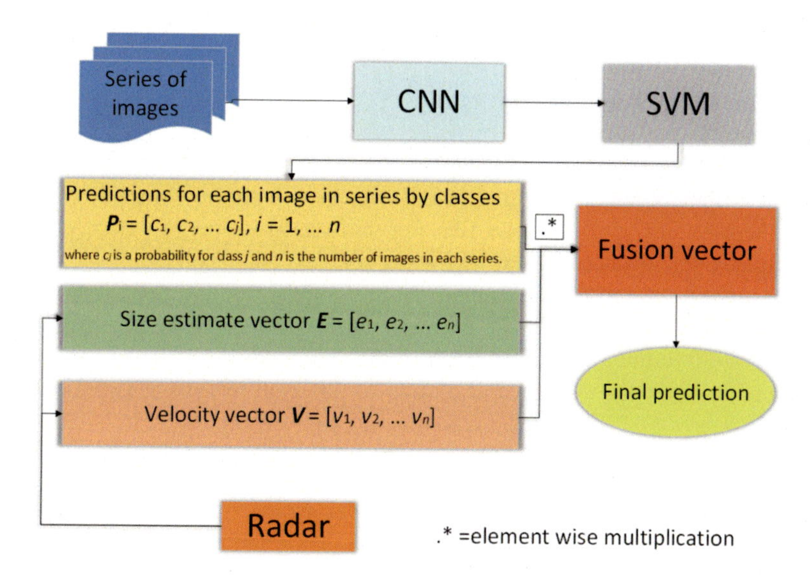

Fig. 7 The classification process

3.3 Fusion

The final classification is achieved by a fusion between the parameters provided by the radar and the predictions from the image classifier. The combined $\boldsymbol{P}$-vector for a series of images is:

$$\boldsymbol{P} = \sum_{i=1}^{n} \boldsymbol{P}_i, \tag{9}$$

where n is the number of images in each series, and the fusion vector, $\boldsymbol{\Phi}$, is:

$$\boldsymbol{\Phi} = \boldsymbol{P} .* \boldsymbol{V} .* \boldsymbol{E} \tag{10}$$

where ".*" denotes element-wise multiplication. The score, S, for final prediction is:

$$S = \mathbf{max}_j(\boldsymbol{\Phi}), \tag{11}$$

$$j = \mathbf{argmax}_j(\boldsymbol{\Phi}), \tag{12}$$

where j is the index of the predicted class.

4 Results

The data augmentation algorithm proved to be slow with our computer resources when the step size, s, is less than 100. For this reason we initially trained several classifier models on the original image set (i.e., no data augmentation) and also several models on various augmented data sets with s as selected from {25, 50, 75, 100, 150, 200, 250, 300, 1000}, respectively. The original data set size was $6 \cdot 631 = 3786$ at this phase. The effect of the data augmentation on classification performance has been examined according to this data set, and it is presented in Fig. 8.

Based on observations of the first training process, it was obvious that to create augmented data sets with all previously applied values of the step size, s, is time-consuming and not reasonable. It was also intuitively clear that the number of epochs should decrease when the number of training examples increases and the learning rate drop period should increase as the number of training examples increases in order to avoid overfitting. Hence we trained the following classifier models with the step side, $s = 50$, when the data augmentation was applied. The classification performance for the latest (i.e., trained on the original data set size at $6984 = 6 \cdot 1164$) models in terms of true positive rate (TPR) is as an average 0.7684

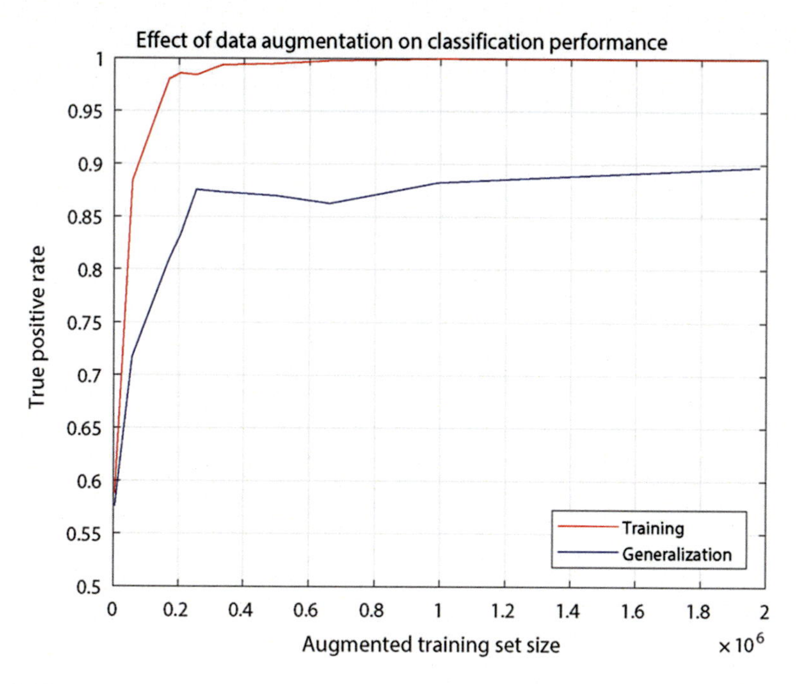

Fig. 8 The upper curve is for testing in training, and the lower curve is according to the generalization test. Several models were trained for each step size, *s*, in the data augmentation. The average TPR values of the models with the same step size are applied in the figure. The starting value for both curves is the average value of the models trained on the first original data set

for the original data set and varies between 0.9317 and 0.9999 for the augmented data sets, which shows clear improvement as augmented training set size increases and especially compared to the models of the original data set.

Generalization was also tested based on the latest models by feeding them images they have never seen. According to this test, the system achieves its state-of-the-art performance of 0.9583 (as TPR) with the augmented data set of the size $1.83*10^6$ (i.e., the color conversion step size 50), number of epochs 8, learning rate drop period (LRDP) 3 with 0.1 as the factor and 0.01 as the initial learning rate, and the dropout applied. See [19] for earlier results.

The results for the original data set (i.e., the data set size $6984 = 6 \cdot 1164$) are presented as a confusion matrix in the Table 1. The results for the best augmented set are presented as a confusion matrix in the Table 2. Cross-validation is applied, and the split into a training set and a validation set was 70% and 30%, respectively, in the two previous data sets. The results for the generalization (the best performed model based on the augmented sets is applied) are presented as a confusion matrix in the Table 3. There were 100 randomly picked images from each 6 classes, which the tested model has never seen before. Thus the test set size in the generalization test was $6 \cdot 100 = 600$. The abbreviated class names, which are names

Table 1 Classification performance of the original data set size at 6984 = 6 · 1164 presented as a confusion matrix

	CG	WTE	HG	LBBG	GC	C/AT	TP/class
CG	245	4	40	28	17	15	70.2%
WTE	0	305	6	6	20	12	87.4%
HG	13	6	249	39	6	36	71.3%
LBBG	14	1	37	252	3	32	72.2%
GC	21	24	5	12	277	10	79.4%
C/AT	6	6	35	16	5	281	80.5%

Table 2 Classification performance of the best augmented data set presented as a confusion matrix

	CG	WTE	HG	LBBG	GC	C/AT	TP/class
CG	91,489	0	0	0	1	0	99.999%
WTE	0	91,485	1	0	3	1	99.995%
HG	0	0	91,450	4	1	35	99.956%
LBBG	0	0	1	91,474	6	9	99.983%
GC	3	12	9	1	91,457	8	99.964%
C/AT	0	0	6	3	2	91,479	99.988%

Table 3 Classification performance tested on 100 unseen images from each class as a confusion matrix

	CG	WTE	HG	LBBG	GC	C/AT	TP/class
CG	97	0	0	0	2	1	97%
WTE	0	99	0	1	0	0	99%
HG	0	1	98	0	0	1	98%
LBBG	0	1	5	87	1	6	87%
GC	0	2	0	1	97	0	97%
C/AT	0	0	2	0	1	97	97%

of the bird species, in the figures are *CG = common goldeneye, WTE = white-tailed eagle, HG = herring gull, LBBG = lesser black-backed gull, GC = great cormorant, and C/AT = common/Arctic tern*. The most right-handed column, *TP/class*, is for correctly predicted images for the class (i.e., species).

5 Discussion and Conclusions

A lot of research is done to monitor birds in wind farms and to find suitable methods for collision detection, e.g. the WT-Bird of the Energy Research Centre of the Netherlands [21, 22]. The principle of the WT-Bird system is that a bird collision is detected by the sound of the impact (triggering) and that the species can be recognized from video images [23, 24]. However, it has known problems with false alarms in high wind circumstances concerning larger bird species, and it has no automated species identification algorithm [25]. We are also aware of two commercial systems: the DTBird of Liquen Consultora Ambiental, S.L., Spain, and the MUSE of DHI, USA [26, 27]. They both promise to detect birds automatically and prevent possible collisions in the vicinity of wind turbines, but no detailed technical data are available. It seems according to their websites that their systems work from an individual wind turbine and they have short distance camera systems for providing images of birds. Only the MUSE seems to have automated identification of species, but the methods are unknown to us. All of the three systems use video footage for detecting possible collisions, and it seems that they have a video camera and recording system for each wind turbine. The primary objective of the system here presented is to use only one camera location for monitoring and species identification, but a video camera (and an infrared camera) is a noteworthy possibility for collision detection. However, we are seeking solution that operates from a single location, which is onshore.

We are currently working on the collision detection problem, but no collisions have been observed until now while the pilot wind turbine has been manually monitored for 18 months. It seems that collisions are quite rare in the research area, and this makes the field testing of the possible collision detection methods challenging.

We proposed a novel system for automatic bird identification as a real-world application. However, the system has restrictions such as the background of the images is the sky (we have not tested images taken otherwise), and images cannot be taken in pitch-dark or in poor visibility conditions. Infrared cameras may contribute to the collision detection, but their contribution to classification is poor because all color information is lost. The proposed system is still under construction and installation phase, so we have not yet been able to test the complete system.

We built and tested several image classifiers based on the small CNN trained with the deep learning algorithm. The best performed classifier proved to be discoverable by changing the step size in the data augmentation and the hyperparameter values (number of epochs and the learning rate drop period) and with or without the dropout technique applied. Our model may be seen as a non-deep or small neural network since it has only three convolution layers.

We demonstrated that the small CNN applied with our data augmentation algorithm for the training data achieved the acceptable state-of-the-art performance of 0.9583 as an image classifier. We showed that our data augmentation method is suitable for image classification problem, and it significantly increases the performance of the classifier.

The measured performance of the image classifier has been obtained without the support of the parameters supplied by the radar. The parameters provide additional and relevant a priori knowledge to the system, and they can turn a misclassified (by images) class into the correct one. We also collect more data continuously at the test site, and thus the number of training examples increases resulting in a better performance of the current classes, and we are able to increase the number of classes as well.

Acknowledgements The authors wish to thank Suomen Hyötytuuli for the financial support and Robin Radar Systems for the technical support with the applied radar system.

References

1. Baxter, A.T., Robinson, A.P.: A comparison of scavenging bird deterrence techniques at UK landfill sites. Int. J. Pest Manag. **53**, pp. 347–356 (2007). Taylor & Francis Online. https://doi.org/10.1080/09670870701421444
2. pelco-D protocol. Bruxy REGNET. http://bruxy.regnet.cz/programming/rs485/pelco-d.pdf
3. All About the Peregrine Falcon. U.S. Fish and Wildlife Service (1999). https://web.archive.org/web/20080416195055/http://www.fws.gov/endangered/recovery/peregrine/QandA.html#fast
4. Statistics. Finnish Meteorological Institute. http://ilmatieteenlaitos.fi/tuulitilastot
5. Robin Radar Models. Robin Radar Systems B.V. https://www.robinradar.com/
6. Richards, M.A.: Fundamentals of Radar Signal Processing. The McGraw-Hill Companies, New York (2005). ISBN: 0-07-144474-2
7. Bruderer, B.: The Study of Bird Migration by Radar, Part1: The Technical Basis. Naturwissenschaften, vol. 84, pp. 1–8. Springer-Verlag, Heidelberg (1997)
8. Fuzzy Logic Toolbox Documentation. The MathWorks Inc. https://se.mathworks.com/help/fuzzy/fuzzy.pdf
9. Mamdani, E.H., Assilian, S.: An experiment in linguistic synthesis with a fuzzy logic controller. Int. J. Man Mach. Stud. **7**(1), 1–13 (1975). Elsevier
10. Goodfellow, I., Bengio, Y., Courville, A.: Deep Learning, pp. 330–371. MIT Press, Cambridge (2016). www.deeplearningbook.org
11. Wang, J., Perez, L.: The Effectiveness of Data Augmentation in Image Classification Using Deep Learning. Stanford University, Stanford (2017). http://cs231n.stanford.edu/reports/2017/pdfs/300.pdf
12. Speranskaya, N.I.: Determination of spectrum color co-ordinates for twenty-seven normal observers. Opt. Spectrosc. **7**, 424–428 (1959). Springer
13. Stiles, W.S., Burch, J.M.: NPL colour-matching investigation: Final report. Opt. Acta **6**, 1–26 (1959). Taylor & Francis
14. Wyszecki, G., Stiles, W.S.: Color Science: Concepts and Methods, Quantitative Data and Formulae, 2nd edn. Wiley, New York (1982)
15. Stockman, A., Sharpe, L.T.: Spectral sensitivities of the middle- and long-wavelength sensitive cones derived from measurements in observers of known genotype. Vis. Res. **40**, 1711–1737 (2000). Elsevier
16. CIE Proceedings, Vienna Session 1963. Committee Report E-1.4.1, vol. B, pp. 209–220. Bureau Central de la CIE, Paris (1964)
17. Blackbody Color Datafile.: Vendian.org. http://www.vendian.org/mncharity/dir3/blackbody/UnstableURLs/bbr_color.html
18. Jarrett, K., Kavukcuoglu, K., Ranzato, M.A., LeCun, Y.: What is the best multi-stage architecture for object recognition. In: International Conference on Computer Vision, pp. 2146–2153. IEEE, Kyoto, Japan (2009)
19. Niemi, J., Tanttu, J.T.: Automatic bird identification for offshore Wind farms: a case study for deep learning. In: Proceedings of ELMAR-2017, 59th IEEE International Symposium ELMAR-2017, Croatian Society Electronics in Marine (2017). ISBN:978-953-184-230-3
20. Huang, J.F., LeCun, Y.: Large-scale learning with SVM and convolutional nets for generic object categorization. In: Computer Vision and Pattern Recognition Conference (CVPR06). IEEE Press, New York, NY (2006)
21. Desholm, M., Kahlert, J.: Avian collision risk at an offshore wind farm. Biol. Lett. **1**, 296–298 (2005). The Royal Society Publishing https://doi.org/10.1098/rsbl.2005.0336

22. Marques, A.T., et al.: Understanding bird collisions at wind farms: an updated review on the causes and possible mitigation strategies. Biol. Conserv. **179**, 40–52 (2014). Elsevier
23. Verhoef, J.P., Westra, C.A., Korterink, H., Curvers, A.: WT-Bird A Novel Bird Impact Detection System. ECN Research centre of the Netherlands (2002). https://www.ecn.nl/docs/library/report/2002/rx02055.pdf
24. Wiggelinkhuizen, E.J., Barhorst, S.A.M., Rademakers, L.W.M.M., den Boon, H.J.: Bird Collision Monitoring System for Multi-Megawatt Wind Turbines, WT-Bird: Prototype Development and Testing. ECN Research Centre of the Netherlands (2006). https://www.ecn.nl/publications/PdfFetch.aspx?nr=ECN-E--06-027
25. Wiggelinkhuizen, E.J., den Boon, H.J.: Monitoring of Bird Collisions in Wind Farm Under Offshore-Like Conditions Using WT-BIRD System: Final Report. ECN Research Centre of the Netherlands (2009). https://www.ecn.nl/docs/library/report/2009/e09033.pdf
26. DTBird.: Liquen Consultora Ambiental, S.L. http://www.dtbird.com/
27. MUSE.: DHI. https://www.dhigroup.com/global/news/2017/02/automated-bird-monitoring-system-lands-on-pioneer-us-wind-farm

Towards an Ecosystem Approach to Assess the Impacts of Marine Renewable Energy

Jean-Philippe Pezy, Aurore Raoux, Nathalie Niquil, and Jean-Claude Dauvin

Abstract Since the beginning of the 2000', the French government ambition was to have an offshore wind production formed 40% of the renewable electricity in 2030. Three calls tenders of Offshore Wind Farms (OWFs) construction have been pronounced since 2011. However, no offshore wind farm (OWF) had been constructed at the end of 2017 due to long administrative procedures and numerous appeals in justice, at French and European levels. Nevertheless, several studies have been enterprised to identify the environmental conditions and ecosystem functioning in selected sites before OWF implantations. However, these studies are generally focused on the conservation of some species or groups of species, and there is no holistic study on the effects of the construction and operation of OWF on an ecosystem taken as a whole. In 2017, a complete and integrated view of the ecosystem of two future OWF sites of the eastern English Channel (Courseulles-sur-Mer and Dieppe-Le Tréport) was developed to model the marine ecosystems before OWF implementation and to simulate reef effects due to new spatial occupation of maritime territory. Results contribute to a better knowledge of the impacts of the OWFs on the functioning of marine ecosystems. They also allow to define recommendations for environmental managers and industry in terms of monitoring the effects of marine renewable energy (MRE), not only locally but also on other sites, at national and European levels.

Keywords Offshore wind farm · Ecosystemic approach · Ecopath with EcoSim · Ecological network analysis · Monitoring · Benthos · Demersal fish · Suprabenthos

---

J.-P. Pezy (✉) · A. Raoux · J.-C. Dauvin
Laboratoire Morphodynamique Continentale et Côtière, CNRS, Normandie Univ, UNICAEN, UNIROUEN, Caen, France
e-mail: jean-philippe.pezy@unicaen.fr; aurore.raoux@unicaen.fr; jean-claude.dauvin@unicaen.fr

N. Niquil
Unité Biologie des Organismes et Ecosystèmes Aquatiques (BOREA), MNHN, CNRS, IRD, Sorbonne Université, Université de Caen Normandie, Université des Antilles, Caen, France
e-mail: nathalie.niquil@unicaen.fr

R. Bispo et al. (eds.), *Wind Energy and Wildlife Impacts*,
https://doi.org/10.1007/978-3-030-05520-2_10

1 Introduction

In the face of global climate change linked to CO2 emissions, there is a worldwide demand for energy transition, and the development of marine renewable energy (MRE) has become a major policy objective [1]. For instance, the European Union (EU) has set a target of 20% of the energetic consumption to be derived from renewable energy sources by 2020 (Directive 2009/28/EC). In line with these political imperative, MRE sector is growing significantly to allow the energetic transition. Among these MRE, offshore wind farm (OWF) has seen consistent growth in capacity, and it is by far the most technically advanced of all MRE [2]. However, this OWF development has raised concerns about their potential impacts on ecosystems [3]. In fact, the construction of OWF is responsible for change in the surrounding environment, and so far, the different ecosystem components (i.e. benthos, fish and birds) that have been studied have already shown some degree of response to OWF. For instance, OWF will be responsible for the introduction of hard substratum [4, 5] which will create new habitats for several benthic and epibenthic species [6]. Thus, OWF may act as artificial reefs due to their foundations and potential scour protections [6, 7]. At the same time, OWF may also include a local habitat loss or degradation [4]. Offshore wind farms (OWFs) that have been built in Europe have been subjected to an environmental impact assessment (EIA) in order to identify both the environmental conditions and ecosystem functioning in selected sites (before OWF construction) and the effects of these new structures on the surrounding marine ecosystems. However, one of the main issues linked to these monitoring programmes is that they are focused on the conservation of some species or groups (iconic or flag species), and there is no holistic study on the effects of OWF construction and operation on an ecosystem taken as a whole [7]. Thus, the environmental impacts of OWF construction at the ecosystem scale remain unclear [3].

In France, no OWF has been yet constructed. However, three successive calls for tenders related to OWF development have been successively published and seven sites have been selected for future OWF construction [8, 9]. Following these calls for tenders, several studies have been enterprised to identify the environmental conditions and ecosystem functioning in selected sites before OWF constructions. In 2017, Raoux et al. [7], Raoux [8], and Pezy [9] explored a new way to look at the potential impacts of OWFs through food web models and flow analysis. In fact, they developed an ecosystem-integrated view of two future OWF sites of the eastern English Channel (Courseulles-sur-Mer and Dieppe-Le Tréport) to describe the marine ecosystems before OWF construction and to simulate reef effect.

In this article, we summarise the present state of the art of the offshore renewable energy development in France and more particularly in the eastern English Channel, with a special emphasis on the issues that are hampering its effective implementation. We also present the methodology and the main results of the MRE ecosystem approach developed by Raoux et al. [7], Raoux [8], and Pezy

[9]. Finally, we also conclude with the lessons learned from this modelling and with the recommendations for future monitoring programmes that we can derive from it.

2 OWF Development State of the Art

2.1 OWFs in France

Renewable energy sources research and development raised a challenge for maritime countries in Europe in the last decades. For more than 15 years, OWFs have been built across Europe [8]. OWF developments were most important in the north European countries (Denmark, Germany, Netherlands, the UK) than in the south European countries like those around the Mediterranean Sea, where the wind farm had been developed in land (Spain, Greece, Italy). With more than 11 million km^2 of waters under its jurisdiction, France holds a huge natural potential for marine renewable energy [8]. At the scale of Europe, France has the second potential of OWF after the UK. The French government has since the beginning of the 2000' the ambition that the offshore wind production formed 40% of the renewable electricity in 2030.

Three successive calls for tenders linked to the offshore wind development had been successively published. The first call for tenders was pronounced on 11 July 2011 and concerned five zones for a total production of 3000 MW: Dieppe-Le Tréport, Fécamp, Courseulles-sur-Mer, Saint-Brieuc and Saint-Nazaire, four of them concerned the English Channel (EC) and three the eastern part of the EC (Fig. 1). The candidates have been selected following three main criteria: the quality of the industrial and social project, the price of the electricity and the respect of the marine environment and the sea users. Only four candidates had been retained: Fécamp (Seine-Maritime, 498 MW), Courseulles-sur-Mer (Calvados, 450 MW), Saint-Nazaire (Loire-Atlantique, 480 MW) and Saint-Brieuc (Côtes d'Armor, 500 MW). After the administrative authorisations and the maritime concessions, the beginning of the production had been estimated to 2020–2021 (10 years of administrative procedures).

The second call was pronounced on 18 March 2013 and concerned two zones: Dieppe-Le Tréport (Seine-Maritime, 500 MW) and between the Yeu and Noirmoutier islands (Loire-Atlantique, 500 MW) near the Loire estuary (Fig. 1).

The third call published in April 2017 concerned only one site offshore of Dunkerque in the southern part of the North Sea (Fig. 1).

At the same time, the French government wants to develop a floating wind turbine sector. A call for projects for the deployment of pre-commercial floating wind farms was launched on 5 August 2015. Four projects were founded of which three are in the Mediterranean Sea and one along the Atlantic coast:

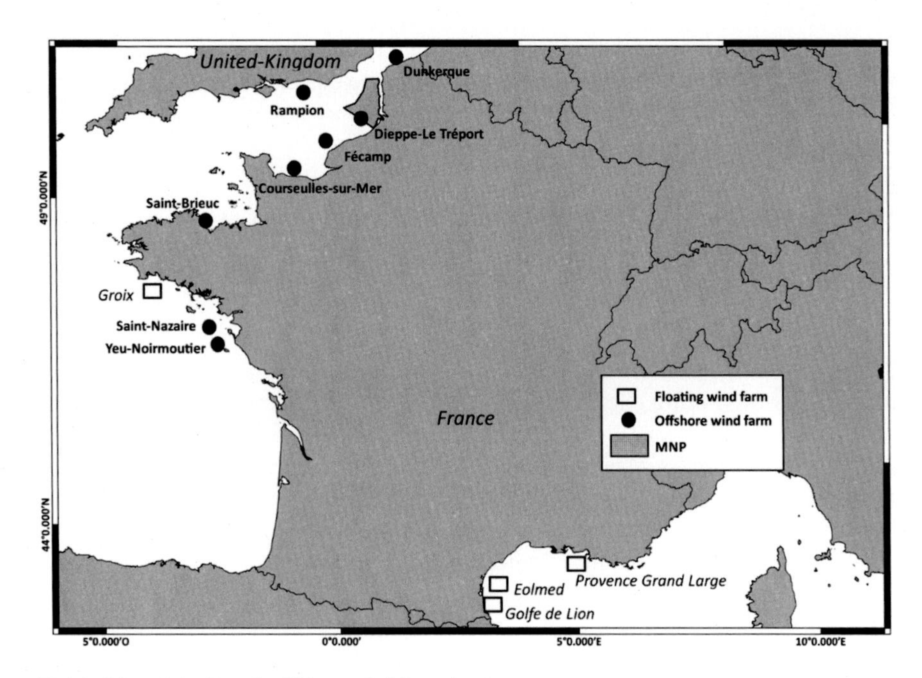

Fig. 1 Map of the French offshore wind farm development

- 'Provence Grand Large' project is owned by 'Electricité De France Energies Nouvelles' (EDF EN) and will comprise three turbines (8 MW).
- 'Eoliennes flottantes Golfe du Lion' project is owned by Engie/EDP Renewables/Caisse des Dépôts and will comprise four turbines (6 MW).
- 'Eolmed' project is owned by Quadran and will comprise four turbines (6 MW).
- 'Eoliennes flottantes de Groix' project is owned by Eolfi/CGN (China Guangdong Nuclear Power Group) and will comprise four turbines (6 MW).

OWF construction follows a regulatory framework based on the maritime public domain (decree n°2004-308 of 29 March 2009) and on the environmental code (article L214-2). Moreover, the General Tax Code (article 519 B) is beneficial for municipalities and sea users, and annual tax on electricity production encourages using mechanical wind energy located offshore (territorial sea). A simplification and consolidation of the offshore wind farm's legal framework was revised in the publication of decree n°2016-9. One improvement, for instance, is that the allowance to occupy a maritime public domain was increased from 30 to 40 years.

A guide to assess the environmental impact of the offshore wind farm construction and operation was published in 2010 and is revised each year [8]. This guide takes into account the evolution of methods and knowledge as well as feedback on the operation of OWF. This document aims to support actors in the offshore wind power branch during the EIA. It is targeted at all actors concerned by the OWF development.

The implementation of this infrastructure is a challenge for developers from technical, legal, social, and environmental points of view. Indeed, these OWFs will be integrated into ecosystems already subjected to a growing number of anthropogenic disturbances such as pollution, transportation, fishing, aggregate extraction or sediment dredging and deposit [7, 10–12]. Nevertheless, 7 years after the first call, no OWF was constructed in France and probably, if they exist, the construction would begin in 2021.

2.2 *OWFs in Eastern English Channel (EC)*

Four projects were raised in the eastern part of the EC, one along the UK and three along France. The following paragraphs present the four OWFs that have been built or will be built in the future in the eastern EC.

The Rampion OWF The Rampion OWF (79 km^2; 13 and 25 km to the coast; 400 MW) was constructed by E.ON off the Sussex coast in the UK. The construction is expected to be completed in 2018. E.ON's final plans use 116 turbines of approximately 3.45 MW capacities. The project was approved by the UK government in July 2014. An onshore construction work began in June 2015 with construction of a new electricity substation adjacent to the existing National Grid Bolney. The first wind turbine was lifted into place in March 2017, with work to backfill the cable duct trenches off Lancing beach initially due to be completed in Spring 2017. Electricity production is expected to start during the third quarter of 2017. Construction of the wind farm is expected to be completed in 2018 at a cost of £1.3 billion (https://www.eonenergy.com/). Thus, the delay between the decision and the electricity production in the case of the UK procedure was about 4 years.

The Courseulles-sur-Mer OWF The project is owned by 'Eoliennes Offshore du Calvados', a subsidiary of Eolien Maritime France and WPD Offshore. The Eolien Maritime France (EMF) was allowed to operate the offshore wind farm off from Courseulles-sur-Mer by the Ministerial Order of April 18 2012. The proposed wind farm will be located 10–16 km offshore from the coast of Calvados, Normandy. The wind farm will have a total area of approximately 50 km^2. It will comprise 75 turbines (6 MW) giving a combinated nameplate capacity of 450 MW. The wind farm turbines will be connected via an interarray network of 33 kV AC cables which will link at one offshore transformer substation located within the wind farm. From this station power will be exported via two 225 kV AC marine cables. The turbines are supported by 7 m of diameter monopiles driven into the sea bed. The footprint of the 75 turbines foundation and of the converter station will be 0.158 km^2 or 0.03% of the overall wind farm area. Our work hypothesis was that scour protections will be installed around the 75 turbines and the converter station and 33% of the cables will be rock-dumped; thus the footprint would amount to 0.342 km^2, or 0.72% of the OWF area. The production generated by the wind park would cover the average annual electricity consumption of approximately 630,000 people, i.e. around 40% of the inhabitants of the surrounding region of Normandy. The commissioning would be provided for 2021.

The Fécamp OWF The project is owned by 'Eoliennes Offshore de Fécamp', a subsidiary of EMF. The proposed wind farm will be located 13–22 km offshore from Fécamp. The wind farm will have a total area of approximately 67 km^2. The wind farm will comprise 83 turbines (6 MW) giving a combined nameplate capacity of 498 MW. The foundations correspond to gravity base structures. The wind farm turbines will be connected via an interarray network of 33 kV AC cables, which will link at one offshore transformer substation located within the wind farm. From this station, power will be exported via two 225 kV AC marine cables. The commissioning would be provided for 2021.

The Dieppe-Le Tréport OWF Following the call of tender launched by the government in March 2013, the Dieppe-Le Tréport offshore wind farm project is owned, on June 3, 2014, by the society 'Eoliennes en mer Dieppe-Le Tréport', a subsidiary of ENGIE, EDP Renewables and Caisse des Dépôts whose exclusive supplier is ADWEN (100% subsidiary of Siemens Gamesa Renewable). The wind farm will be located 15.5 km offshore Le Tréport and 17 km offshore Dieppe, in water depths ranging from 12 to 25 m. The project will cover an area of approximately 83 km^2, with a total of 62 turbines, each having a capacity of 8 MW, giving a combined installed capacity of 496 MW. The wind farm turbines will be connected via an interarray network of 33 kV AC cables, which will link at one offshore transformer substation located within the wind farm. From this station, power will be exported via two 225 kV AC marine cables. The foundations correspond to jacket structures. The turbines would be spaced of 1000 m in order to favour the fishing. The commissioning would be provided for 2021.

2.3 *Issue Hampering MRE Development*

The French government wants to develop OWF activities along its coasts. However, this development is hampered by too many procedures and actions for annulment brought by fisherman and environmental protection associations. For instance, the public inquiry, who had been appointed by the Seine-Maritime Prefect, gave a favourable opinion on the construction of the Fécamp OWF on December 1, 2015. However, today the project is still hampered by actions for annulment brought by fisherman on June 13, 2017 who have lodged complaints to the European Commission.

On the same line, the Marine Nature Park 'Estuaires Picards et mer d'Opale' (MNP) management board has not given a favourable opinion concerning the Dieppe-Le Tréport OWF construction on October 20, 2017. This decision was reached after careful analysis of the environmental impact assessment. However, it is worth to know that this decision is no longer decisive but advisory. Indeed, the final decision belongs to the French Agency for Biodiversity which gave a favourable opinion on February 20, 2018.

In front of the long delay to the implantation of OWF along the French metropolitan coast, at the end of November 2017, a part of the discourse of the Prime Ministry Edouard Philippe at the 'Assises de l'économie de la mer', Le Havre, concerned the OWF development in France: 'Wind turbines at sea in France; it's mostly for the delay. A delay which, in the country of the Rance tidal plant, made a little mess. A delay that we tried to catch up in 2015. As soon as 2018, we will launch the preliminary studies for the application of future tenders on wind power floating in Brittany and the Mediterranean. We will also launch environmental studies and public debate on the draft wind laid off the coast of Oléron. To better manage future conflicts of use, I have asked the prefects to implement strategic planning and space of the maritime spaces and their copy by summer 2018. I asked to the Environment ministry, a radical simplification of procedures of education work. The goal? Be traced back as early as possible the preliminary, in particular environmental studies. And make them drive by the State. Of course, it is not to deny conflicts of use. The sea belongs to anyone. It belongs to all. But a conflict, it is anticipated, it is defused, it can subside too, thanks to studies of quality and a serious dialogue'.

In addition, the administrative procedures of the French government are not always consistent between the different sites of OWF construction leading to different monitoring strategies. Thus, monitoring surveys will differ from one site to another. This is the case for the Courseulles-sur-mer and the Fécamp OWF sites which are located in the same region (Normandy) and are situated some 100 km from one another.

For the Courseulles-sur-Mer OWF project, a monitoring committee and a scientific committee are scheduled to be set up under the authority of the Calvados department prefect and maritime prefect. This committee would be composed by the regional fisheries committee, the local communities and scientists. The offshore wind farm consortium will present the environmental monitoring strategy to the committee which will be in charge to analyse and validate it. For the Fécamp offshore wind farm site, a scientific monitoring committee was created under the authority of the department prefect and maritime prefect. This scientific committee relies on the Commission Régionale Mer et Littoral (COMEL), extended to the French Research Institute for the Exploitation of the Sea (Institut Français de Recherche pour l'Exploitation de la MER, IFREMER) and the French Agency for Biodiversity (AFB). The monitoring studies should be also presented to the scientific committee for recommendation before validation.

In the end, pragmatism prevailed in the UK procedures which led to the construction of the Rampion offshore wind farm in 4 years, whereas no OWF has been yet constructed (in 7 years) along the French coasts due to many complex procedures. There is an urgent need to simplify the French procedures. One way would be to have only one operator responsible for the coordination of the OWF construction environmental studies along the French coasts as found in Belgium. For instance, each year the Royal Belgium Institute of Natural Sciences publishes the main scientific findings of the Belgian offshore wind farm monitoring programme [13]. These monitoring programmes target physical (i.e. hydro-geomorphology), biological (i.e. fish communities, soft substrate macrobenthos, fish communities, hard substrate) and socioeconomic aspects [13].

3 An Ecosystem Approach for Dieppe-Le Tréport and Courseulles-sur-Mer Wind Farm Sites

3.1 The Ecopath Modelling Approach

In order to gain further knowledge on the ecosystem structure and functioning before the construction of the Dieppe-Le Tréport (DLT) and the Courseulles-sur-Mer OWF, trophic web models describing the initial state of the both sites were built by Pezy [9] and Raoux et al. [7]. They used the software Ecopath with EcoSim (EwE) [14–16] to model the trophic web at the two sites. This approach, in which all biotic components of the system could be considered at the same time, is useful to gain a better understanding of the system trophic structure and functioning and for predicting how it may change over time when facing perturbations [17]. The proposed model considered the full range of size classes of biota, from prokaryotes to top predators. In addition, this approach also allows to provide measure of the ecosystem emergent properties through the calculation of Ecological Network Analysis indices [18]. More details are given in Raoux et al. [7].

Dieppe-Le Tréport OWF Site As mentioned below, Pezy [8] built an Ecopath model to describe the initial state of the Dieppe-Le Tréport OWF site. For that the collection of new data on biological compartments (zooplankton, suprabenthos, meiofauna, benthos and demersal fishes) was essential. Four cruises (summers 2014/2015 and winters 2015/2016) allowed to estimate the contribution of each zoological group in three main benthic habitats, i.e. sandy gravels (sG), gravelly sand (gS) and medium clean sand (S), founded on the site, corresponding to two benthic communities well represented in the EC (Fig. 2). The zooplankton were sampled with WP2 (200 μm) nets in two stations from two benthic habitats (sG and S) with day and night samplings. Benthic macrofauna was sampled with a Van Veen grab (0.1 m^2) at 25 stations (with 5 replicates) (Fig. 2). The suprabenthos was sampled with a modified Macer-Giroq sledge [19] in two stations from two benthic habitats (sG and S) with day and night samplings (Fig. 2). Demersal fish was sampled with a beam trawl (3 m) in ten stations. Stomach content analysis of demersal fish was carried out (Fig. 2). In fact, this analysis allowed to quantify the contribution of benthic preys located in the future implementation site of the offshore wind farm to the diet of demersal fish species and so to identify the benthic species playing a key role in the trophic web. The models were composed by 27 compartments from bacteria to cetacean.

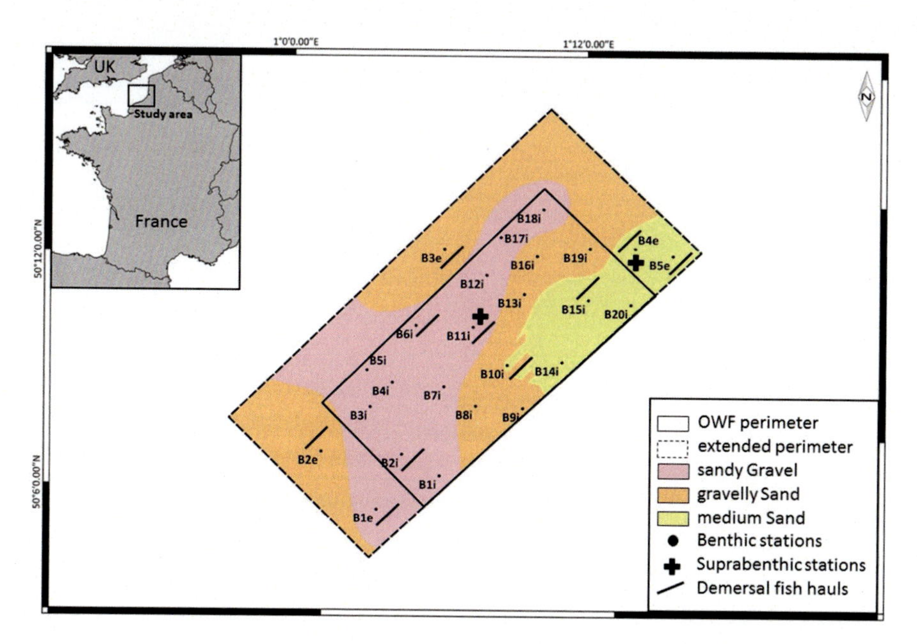

Fig. 2 Localisation of the different compartment samples (benthos, suprabenthos and demersal fish) at the Dieppe-Le Tréport site

3.1.1 Courseulles-sur-Mer OWF Site

On the same line, Raoux et al. [7] ran also the EwE approach to model the trophic web at the site of the construction of the future Courseulles-sur-Mer OWF. Thus an Ecopath ecosystem model composed of 37 compartments, from phytoplankton to seabirds, was built to describe the situation 'before' the OWF construction. It was built by using the data collected during the EIA of the Courseulles-sur-Mer (CSM) wind farm site.

The Dieppe-Le Tréport and the Courseulles-sur-Mer OWFs share in common a same sedimentary type which is the gravelly sand. The ecosystem biomasses of the gravelly sand of the both sites (DLT and CSM) are dominated by benthic invertebrate filter feeders with a biomass of 45 gC.m^{-2} at the DLT site (dominated by the bivalves *Glycymeris glycymeris* and *Polititapes rhomboides*) and a biomass of 22.5 gC.m^{-2} at the CSM site (dominated by the bivalve *Polititapes rhomboides*) (Table 1).

3.2 Simulation the 'Reef Effect' Due to the Wind Farm Implantation Using EcoSim Simulations

As mentioned before, so far, there is no holistic study on the OWF construction and operation effects on an ecosystem taken as a whole. Indeed, EIA have onlybeen

Table 1 Biomass (gC.m^{-2}) of the different compartments of the two sites

Functional groups	Biomass (gC.m^{-2})	
	DLT	CSM
Fish piscivorous	0.02	0.28
Fish planktivorous	0.62	1.16
Fish benthos feeders	0.007	3.16
Fish flatfish	0.02	0.08
Benthic invertebrate, predators	1.08	2.93
Benthic invertebrate, filter feeders	46.49	23.42
Benthic invertebrate, deposit feeders	4.86	3.54
Meiofauna	1.67	0.97
Suprabenthos	0.56	2.00
Zooplankton	0.61	1.72
Phytoplankton	3.10	3.24
Bacteria	0.75	0.75

DLT Dieppe-Le Tréport, *CSM* Courseulles-sur-Mer

investigated on benthic and fish species, but no studies have adopted a holistic approach to assess its potential impacts on the ecosystem structure and functioning [7, 8]. Recently, Raoux et al. [7] explored a new way to look at the potential impacts of OWFs through food web models and flow analysis. In fact, Raoux et al. [7] derived an EcoSim model (the temporal dynamic module of EwE) to project over the next 30 years the ecosystem evolution after the simulated increase in biomass of some targeted benthic and fish compartments in relation to the wind farm construction (i.e. forcing a reef effect). ENA indices [18] were finally calculated, summarising ecosystem functional traits and giving indications about the possible ecosystem state evolutions at the end of the simulation. Among the core conclusions, this modelling approach showed (1) that the total ecosystem activity, the overall system omnivory (proportion of generalist feeders in the system) and the recycling should increase after the construction of the OWF [7] and (2) that important higher trophic levels such as exploited piscivorous fish species, endangered marine mammals and seabirds might very likely and positively respond to the aggregation of biomass on piles and turbine scour protections. More details were given in Raoux et al. [7].

4 Lessons Learned and Recommendation for the Future

Actually, the EIA considers the sensitivity to potential disruptions of some groups of valuable species without taking into account the links between them [7, 8]. Thus, OWF construction effects on the ecosystem structure and function remain unclear. A more holistic view is needed to allow the passage of the split vision (site by site, activity by activity, country by country) to an ecosystem approach of the EC (especially since the species impacted by OWFs are species with large distribution area (i.e. marine mammals and sea birds)) as recommended by the Marine Framework Directive [12]. The holistic view of OWF impacts on the trophic web developed by Raoux et al. [7] through trophic web modelling could be replicated on other site in the EC (French and English sides) and could be useful to analyse the reef and reserve long-term effects in the context of climate change. Indeed, using quantitative modelling to assess offshore wind farm impacts on the whole ecosystem would allow to bring new knowledge to policy makers. It would also allow for a better integration of ecological considerations in managing decisions and for planning maritime space.

The vision for an ecosystem-based management for both sides of the EC requires a common vision between France and the UK to adapt the future management of this marine space as climate change and human pressures increase (e.g. granulate extraction and new offshore energies, such as wind farms and hydraulic systems). In the future, studies of the cumulative effects of human pressure [10, 11] must be promoted at the level of both the subdivisions proposed, which appears to be a coherent level of ecosystem-based management. However, to date each state manages its own side independently. Thus, France and the UK must imagine synergies to jointly manage this maritime area in a Brexit context.

Acknowledgements These works were co-funded by the Normandy Region, the group 'Eoliennes Offshore du Calvados' and the group 'Eoliennes en mer Dieppe-Le Tréport'. We also acknowledge, for their help in the data sampling, the captain and the crew of the Oceanographic Vessel 'Celtic Warrior'. We acknowledge the two anonymous reviewers for their comments on this article.

References

1. Bergström, L., Sundqvist, F., Bergström, U.: Effects of an offshore wind farm on temporal and spatial patterns in the demersal fish community. Mar. Ecol. Prog. Ser. **485**, 199–210 (2013)
2. Willsteed, E.-D., Jude, S., Gill, A., Birchenough, S.N.R.: Obligations and aspirations: a critical evaluation of offshore wind farm cumulative impact assessments. Renew. Sust. Energ. Rev. **82**, 2332–2345 (2018). https://doi.org/10.1016/j.rser.2017.08.079
3. Bailey, H., Brookes, K.L., Thompson, M.: Assessing environmental impacts of offshore wind farms: lessons learned and recommendations for the future. Aquat. Biosyst. **10**, 1–13 (2014). https://doi.org/10.1186/2046-9063-10-8
4. Petersen, J.K., Malm, T.: Offshore windmill farms: threats to or possibilities for the marine environment. Ambio. **35**, 75–80 (2006). www.jstor.org/stable/4315689
5. De Mesel, I., Kerckhof, F., Norro, A., Rumes, B., Degraer, S.: Succession and seasonal dynamics of the epifauna community on offshore wind farm foundations and their role as stepping stones for non-indigenous species. Hydrobiologia. **756**, 37–50 (2015)
6. Wilhelmsson, D., Malm, T.: Fouling assemblages on offshore wind power plants and adjacent substrata. Estuar. Coast. Shelf Sci. **79**, 459–466 (2008). https://doi.org/10.1016/j.ecss.2008.04.020
7. Raoux, A., Tecchio, S., Pezy, J.-P., Degraer, S., Wilhelmsson, D., Cachera, M., Ernande, B., Lassalle, G., Leguen, C., Grangeré, K., Le loch, F., Dauvin, J.-C., Niquil, N.: Benthic and fish aggregation inside an offshore wind farm: which effects on the trophic web functioning? Ecol. Indic. **72**, 33–46 (2017). https://doi.org/10.1016/j.ecolind.2016.07.037

8. Raoux, A.: Approche écosystémique des Energies Marines Renouvelables: étude de l'impact sur le réseau trophique de la construction du parc éolien au large de Courseulles-sur-Mer et du cumul d'impacts. Thèse de Doctorat, Université de Caen Normandie, France, 293 pp (2017)
9. Pezy, J-P.: Approche écosystémique d'un futur parc éolien en Manche orientale: exemple du site de Dieppe – Le Tréport. Thèse de Doctorat, Université de Caen Normandie, France, 324 pp (2017)
10. Halpern, B.S., Walbridge, S., Selkoe, K.A., Kappel, C.V., Micheli, F., D'Agrosa, C., Bruno, J.F., Casey, K.S., Ebert, C., Fox, H.E., Fujita, R., Heinemann, D., Lenihan, H.S., Madin, E.M.P., Perry, M.T., Selig, E.R., Spalding, M.D., Steneck, R., Watson, R.: A global map of human impact on marine ecosystems. Science. **319**, 948–952 (2008). https://doi.org/10.1126/science.1149345
11. Halpern, B.S., McLeod, K.L., Rosenberg, A.A., Crowder, L.B.: Managing for cumulative impacts in ecosystem-based management through ocean zoning. Ocean Coast. Manag. **51**, 203–211 (2008). https://doi.org/10.1016/j.ocecoaman.2007.08.002
12. Dauvin, J.-C.: Are the eastern and western basins of the English Channel two separate ecosystems? Mar. Pollut. Bull. **64**, 463–471 (2012). https://doi.org/10.1016/j.marpolbul.2011.12.010
13. Degraer, S., Brabant, R., Rumes, B., Vigin, L.: Environmental Impacts of Offshore Wind Farms in the Belgium Part of the North Sea: A Continued Move Towards Integration and Quantification. Royal Belgium Institute of Ntural Sciences, OD Natural Environment, Marine Ecology and Management Section., 141 pp, Brussels (2017)
14. Polovina, J.-J.: Model of a coral reef ecosystem. The ECOPATH model and its application to French Frigate Shoals. Coral Reefs. **3**, 1–11 (1984)
15. Christensen, V., Walters, C.J.: Ecopath with Ecosim: methods, capabilities and limitations. Ecol. Model. **172**, 109–139 (2004). https://doi.org/10.1016/j.ecolmodel.2003.09.003
16. Christensen, V., Walters, C.J., Pauly, D., Forrest, R.: Ecopath with Ecosim version 6 User Guide. Lensfest Ocean Futur. Proj. **2008**, 1–235 (2008)
17. Plagànyi, E.E.: Models for an Ecosystem Approach to Fisheries FAO Fisheries Technical Paper, vol. 477. FAO, Rome. 108 pp (2007)
18. Ulanowicz, R.E.: Growth and Development: Ecosystems Phenomenology. SpringerVerlag, New York. 166 pp (1986)
19. Dauvin, J.C., Lorgeré, J.C.: Modifications du traîneau MACER-GIROQ pour l'amélioration de l'échantillonnage quantitatif étagé de la faune suprabenthique. J. Rech. Océanogr. **14**, 65–67 (1989)

Camera-Trapping Versus Conventional Methodology in the Assessment of Carcass Persistence for Fatality Estimation at Wind Farms

Luís Rosa, Tiago Neves, Diana Vieira, and Miguel Mascarenhas

Abstract In the last decades, there has been a worldwide increase in wind energy. Despite its advantages, wind farms carry negative impacts on bird and bat populations, such as direct mortality due to collision with wind turbines.

Carcass searches beneath the turbines are mandatory to access this impact, as well as the assessment of two correction factors: searcher efficiency and probability of carcass persistence. The latter considers the possibility of carcass removal by scavengers (or any other event such as decomposition) between monitoring sessions, influencing the number of carcasses detected.

Carcass persistence trials consist in randomly placing carcasses under the turbines and, in this study, checking them daily during a 15-day period. Camera traps are looked as an alternative that might reduce human and financial effort while allowing the collection of the exact removal time and characterization of the scavenger's guild.

We conducted trials in three different wind farms, in a total of seven campaigns (15-day carcass checking periods), across different seasons.

We compared the camera-trapping with the conventional methods and analyzed the influence of using continuous vs. censored data on the correction factor estimate and have not found significant differences.

Camera traps allowed the recording of the exact removal time and the identification of the removal agent for most of the carcasses and allowed a significant reduction of the field work and the costs associated.

We present a few guidelines to be taken into consideration when using this method.

Camera-trapping demonstrated to be a good method to replace the conventional method, ensuring at least the same results, while allowing the characterization of the scavenger's guild and a significant cost reduction.

L. Rosa (✉) · T. Neves · D. Vieira · M. Mascarenhas
Bioinsight, Odivelas, Portugal
e-mail: luis.r@bioinsight.pt; tiago.n@bioinsight.pt; miguel.m@bioinsight.pt

R. Bispo et al. (eds.), *Wind Energy and Wildlife Impacts*,
https://doi.org/10.1007/978-3-030-05520-2_11

Keywords Bird fatalities · Bat fatalities · Carcass removal · Camera-trapping · Scavengers

1 Introduction

1.1 Context

Wind power is one of the fastest growing renewable energy sources worldwide [1] and, at the same time, an environmentally friendly source of energy, as it is relatively free of carbon emissions and might contribute to the efforts to counteract climate change caused by increasing human energy needs [2]. Despite wind energy advantages, wind farms carry negative impacts as well, mainly on bird and bat populations [3–8], being one of these impacts the direct mortality or injury due to collision with turbines or barotrauma in bats case [9, 10].

During the operational phase, depending on the characteristics and locations of the wind farm, it is usually mandatory for wind energy developers to implement impact monitoring programs for birds and bats [11], especially on the first years of the wind farm operation. The duration of those programs in Portugal is usually recommended to be 1 year in the preconstruction phase and 3 years in the operational phase, and in some cases, namely, when wind farms are constructed in protected areas, the monitoring may be required for the entire life cycle of the enterprise [11].

To determine the extent of the impact from collision on bird and bat populations, carcass searches near turbines are usually mandatory, to assess the number of fatalities that occur during the operational phase of a wind farm. Only a fraction of the animals that collide with the turbines are effectively found during carcass searches, as some might be removed by scavengers or can disappear due to other events (e.g., decomposition) between searches [12, 13] or might remain undetected due to several factors, e.g., visibility and vegetation density. Therefore, the real mortality must be estimated by correcting the number of observed carcasses with two correction factors: the probability of carcass persistence and the searcher efficiency [14–17].

It is important to assess the probability of carcass persistence and searcher efficiency, as poorly outlined trials might result in heavily biased mortality estimates [15–22].

1.2 Correction Factor Calculation Trials

The searcher efficiency is affected by several factors that can influence the detection rate by a human searcher, such as the vegetation density, carcass size, accessibility of the terrain, weather, state of decomposition, and searcher intrinsic detection ability, which can be different from one observer to another. So, the experimental design of searcher efficiency trials usually integrates several observers and different searching scenarios representative of the wind farm characteristics.

The probability of carcass persistence is obtained through trials, as part of the monitoring programs, where a predetermined number of carcasses of different size classes are distributed randomly in the wind farm, over a defined period. The time until removal (when the carcass or any other evidence of mortality is no longer present) of each carcass is recorded. The number of placed carcasses varies with the size and characteristics of the wind farm. Bird carcasses of three size classes (small, ≤15 cm; medium, 20–25 cm; large, ≥25 cm) are commonly used, as well as mice as surrogates for bats, but vary with the protocol demanded by the local environmental authorities, in order to reflect as faithfully as possible natural removal rates. A minimum distance between the carcasses is also considered when they are distributed, to avoid their presence as an artificial attraction for predators that, in turn, could influence the speed at which carcass is removed [23, 24].

Some studies reinforce that carcass persistence trial results are always site-specific and the permanence probability estimated for a specific wind farm should not be used to correct the observed mortality at another site [25]. Therefore, experimental design must consider factors that might influence the removal rate, such as the season, carcass size, and weather condition [18, 22, 25–27].

It is usually recommended that the carcass persistence trial is planned for several seasons of the year. The time during which the trials are carried out varies. In the present study, the trials were carried out as recommended by the local environmental authorities [28], during a 15-day period, a period that is based on studies in Portuguese wind farms, suggesting that this is the minimum period in which the tests should be carried out in wind farms with mountainous characteristics such as those of the present study [29].

In the conventional methodology, a technician checks each carcass every day during the 15-day period and verifies the presence/absence of it. The persistence time of the known carcass falls between the intervals of time between two inspections, being denominated as interval-censored [25]. This methodology involves a considerable effort, namely, regarding human resources, stays, food, fuel, and tolls.

1.3 Camera-Trapping

The camera-trapping technology in the assessment of probability of carcass persistence has been highlighted by some recent studies [26, 30–34]. This technology has been looked at as an alternative to the conventional methodology used in carcass persistence trials, as it is a less demanding method, in terms of work effort and financial cost, since less frequent visits to the carcasses are required [18, 26, 31], and it also overcomes some shortcomings of the conventional methodology as, in the conventional methodology, only censored data are registered. Some studies have indicated that slight differences in persistence rates can pronouncedly bias mortality estimates [25, 29].

Additionally, it allows to capture the precise moment when a carcass is removed and to identify the agent responsible for the removal [26], which might shed new light on the removal assessment. In some countries, this methodology is already recommended by authorities, e.g., in Scotland [35].

2 Methods

2.1 Study Area

The carcass persistence trials were conducted in three wind farms (identified here as A, B, and C) located at the central and north region of Portugal, all of them located in a series of mountain ridges.

Wind farm A has four turbines installed in an area of high habitat diversity and phytogeographic importance. In this area, there is a contrast between the north and west slopes, with Atlantic characteristics, and the south and east with Mediterranean characteristics. The study area vegetation is dominated by the presence of cluster pine (*Pinus pinaster*) and the presence of *Quercus ilex*, *Quercus suber*, *Castanea sativa*, and other hardwood trees.

Wind farm B has 11 turbines installed in the summit of Lousã Ridge and includes a 60 kV high-voltage power line that connects the wind farm to another wind farm and a substation. The vegetation consists in low shrubs (*Erica* sp. or *Calluna vulgaris*), herbaceous vegetation, oak forests, and plantations of coniferous and mixed woods.

Wind farm C is located in Viseu District and consists of 12 turbines, and it includes a 60 kV high-voltage power line that connects the wind farm to a substation. The vegetation is dominated by scrubs and plantations of coniferous and the presence of plantations of *Eucalyptus*, mixed woods, and agricultural fields.

2.2 Field Methods

The number of trials in each wind farm was defined by their obligatory monitoring protocols, imposed by the authorities (Table 1). Wind farm A trials were conducted from December 08 to 22, 2016. Wind farm B trials were conducted from June 04 to 22and from October 06 to 21, 2016. Wind farm C trials were conducted from January 08 to 27, March 31 to April 13, August 17 to 31, and October 19 to November 02, 2017.

Table 1 Seasons covered by the carcass persistence trials in the different wind farms (winter, December/January; spring, March/April; summer, June to August; autumn, October/November)

Wind farm	Winter	Spring	Summer	Autumn
Wind farm A	✓			
Wind farm B			✓	✓
Wind farm C	✓	✓	✓	✓
Total trials	2	1	2	2

Table 2 Number of carcasses of each size class used in each trial in each wind farm

	Carcasses		
Wind farm	Large	Medium	Small
Wind farm A	7	7	10
Wind farm B	10	10	10
Wind farm C	10	10	10

The number of carcasses of each size class placed per wind farm varied slightly (Table 2) and followed the monitoring protocol designs defined by the authorities, which are based on the wind farm size and characteristics. The total number of carcasses placed underneath the turbines was 204.

We used fresh carcasses of common parakeets (*Melopsittacus undulatus*), quails (*Coturnix coturnix*), and partridges (*Alectoris rufa*) as surrogates for bird carcasses of three size classes: small, $\leq$15 cm; medium, 20–25 cm; large, $\geq$25 cm. The species used were defined by the authorities and are the ones most frequently used in carcass persistence trials in Portugal. Bird carcasses were purchased in avian breeding facilities.

The carcasses were randomly placed, independent of the size class, in the wind farm study areas and their vicinities, as long as the habitat was representative of the wind farm area. Their placement location was randomly defined in GIS environment and classified according to the habitat type (no vegetation, low scrub, high scrub, forest). The distance to the nearest wind turbine was also recorded. The handling of the corpses was performed with lab gloves to prevent human odor contamination. To avoid scavenger swamping, i.e., the saturation of food availability and an abnormal attraction of scavengers, which could bias the results [31], the carcasses were displayed at a minimum distance of 500 m from each other. They were also displayed at a minimum distance of 20 m from roads, to avoid the attraction of scavengers to an eventual roadkill.

Once placed, each carcass was monitored for a period of 15 days, using camera traps. Carcasses were considered removed when there were no remains in the camera field of view and/or the number of feathers present was less than 10 [36]. If an entire carcass or remains stood until the last day of the monitored period, it was considered as a right-censored observation.

We used the infrared camera trap models *HCO Scoutguard SG565*, *LTL Acorn 5210A*, and *Reconyx RC60 Covert IR*, equipped with 2 or 4 Gb *Secure Digital (SD®) memory cards*. The cameras were programmed to take bursts of three pictures when the infrared sensor was triggered. To minimize the risk of filling the memory card because of unwanted trigger of the sensor, related to the movement of surrounding plants or meteorological events, it was defined a 5-min interval between two consecutive bursts, and the sensor sensibility was set to medium. The *LTL Acorn* and *Reconyx* models were also configured to take a picture every 12 h, to assure at least two daily pictures, in case the cameras malfunctioned and were not triggered by the presence of a scavenger.

Every camera was visited 7 days after their placement, for battery and memory card maintenance [31], and collected at the end of the 15 days.

The cameras were attached to a tree, or scrub when the area had no trees, at an angle that would try to avoid direct sunlight on the camera. Carcasses were centered in the camera's field of view, at 1–2 m of distance from the camera.

The conventional method data was simulated from the data obtained through camera-trapping, transforming "exact time of removal" data (not censored) into presence/removal of the carcass in that day (censored data), as would be if the carcasses were visited every day.

2.3 Data Analysis

To evaluate the removal curves for the carcasses, considering the variables *size* and *season*, it was used the *Carcass Persistence* module of the online Wildlife Fatality Estimator (www.wildlifefatalityestimator.com). This module is based on survival analysis techniques [37], since it is intended to analyze *lifetime* data, that is, the time until the occurrence of a certain event, in this specific case, until the removal of the corpse [25, 38].

Once the empirical survival curves $\hat{S}(t)$ were estimated according to the nonparametric model (Kaplan-Meier curves), the parametric model (exponential, Weibull, lognormal, or log-logistic) that presented the best fit to the probability of corpse removal over time was selected. Graphical analysis and Akaike's information criterion (AIC) values were used to compare the models' relative goodness of fit. AIC differences were determined as they reflect the loss of information when the fitted model is used to the detriment of the best-adjusted model [39]. The model with the best relative fit was chosen according to the lowest AIC value and a difference equal to zero. The final model was defined according to a stepwise process, and the nonsignificant variables were excluded from the model ($p > 0.05$).

Once the final model was defined, the correction factors associated with the removal of carcasses were determined according to the estimators of Huso [40] and Korner-Nievergelt et al. [21], respectively:

- Average carcass persistence time ($\bar{t}$) – the average number of days a corpse remains on the ground until it is removed or fully decomposed
- Carcass persistence probability (s) – the average probability of a carcass not being removed in a 24-h period, considering the time interval (I, in days) between prospects ($I = 7$ and/or $I = 28$)

Differences between the removal factor values obtained by the camera-trapping method and the conventional method were analyzed via generalized linear mixed models (GLMM) with a lognormal distributional family. Model residuals were checked, and no significant deviations from the assumed distribution were found. Carcass size and method were used as explanatory variables, and the corresponding second-order interaction was considered. The estimators (Huso and Korner for 7-days and 28-days interval searches) were used as a random factor. Statistical analyses were implemented in R [41], using the lme4 package [42]. Data were analyzed under a 0.05 significance level.

3 Results

3.1 *Carcass Persistence Rates: Camera-Trapping vs. Conventional Method*

The obtained values for both carcass persistence correction factors using the conventional method or the camera-trapping method were not significantly different and did not influence the final fatality estimation (Table 3). Furthermore, neither carcass size nor the interaction between both factors was significant (Table 3).

3.2 *Camera Trap Performance*

The total carcasses placed were 204, from which 156 were removed. It was possible to capture the exact moment of the removal 109 times of them (69.87% of the total carcasses removed) (Fig. 1), and the agent responsible for the removal was identified 98 times (Fig. 2). From those, the fox (*Vulpes vulpes*) and the beech marten (*Martes foina*) were the main responsible for the carcass removal (42.9% and 21.4%, respectively) (Fig. 2). The remaining cases in which the exact removal moment was

Table 3 Summary of GLMM results to examine differences in carcass persistence correction factors

Predictor variable	Estimate	SE	t-value	*p*-value
Intercept	0.572	0.576	0.993	0.321
Camera-trapping	0.030	0.092	0.324	0.746
Medium size	−0.130	0.100	−1.304	0.192
Small size	−0.026	0.094	−0.272	0.786
Camera-trapping: medium size	0.010	0.139	0.069	0.945
Camera-trapping: small size	−0.002	0.132	−0.013	0.990

SE standard error

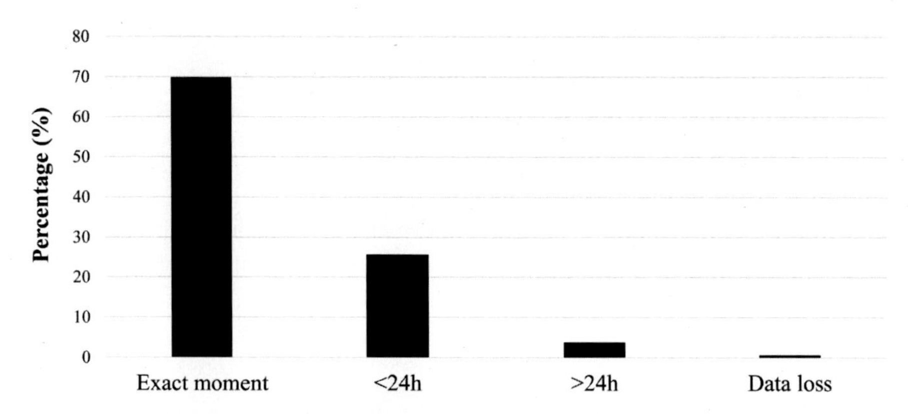

Fig. 1 Camera performance of the total placed cameras, when the carcass was removed, in %

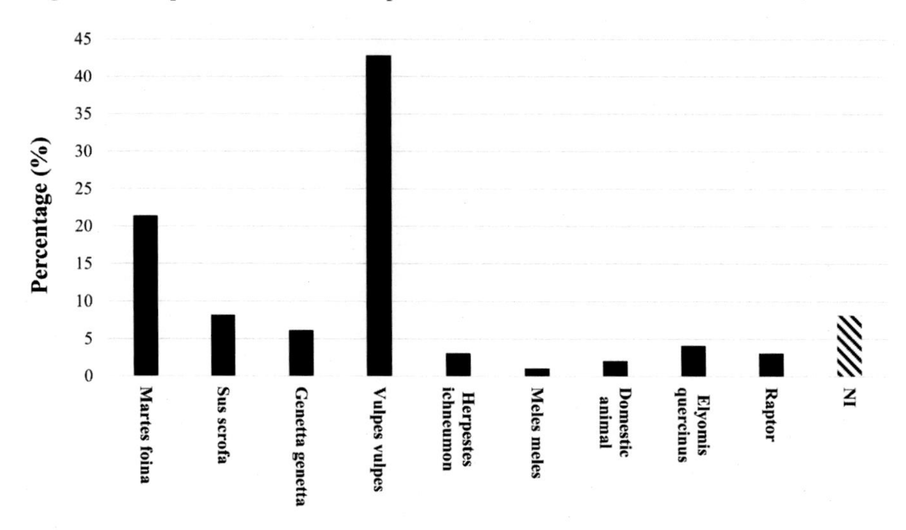

Fig. 2 Species responsible for the carcass removal in the three different wind farms in % (*NI* non-identified)

not captured were divided by removal occurring between two consecutive photos with an elapsed time of less than 24 h (25.64%) and removal occurring between two consecutive photos with an elapsed time of more than 24 h (3.85%) and one stolen camera (0.64%) (Fig. 1).

3.3 Cost Comparison

We conducted an overall cost comparison for both methods (conventional and camera-trapping), to analyze the cost-benefit relation of each methodology (Table 4). In the conventional method, only one technician is required for the work, costing

Table 4 Average expense units per trial, for the three wind farms, overall saving for the seven trials and post-amortization

Expense	Conventional Units (average per trial)	Camera-trapping Units (average per trial)	Saving
Technicians (workdays)	15	7	53%
Stays	14	3	79%
Meals	15	5	67%
Fuel (km)	1200	920	23%
Tolls	2	6	−200%
Carcasses	24	24	0
Camera traps	–	31/7	–
Total saving for the 7 trials		17.5%	
Average saving per trial, after amortization		42.16%	

15 workdays. With the camera-trapping method, there is a first 2-day visit where two technicians are required, followed by a 2- and a 1-day visits where only one technician is required, totaling seven workdays. We also considered the purchase of 31 camera traps. We divided the cost by the seven trials where they were effectively used in this study.

The comparison was between the effective costs of the camera-trapping method and the estimate costs of the conventional method, being considered the prices per unit practiced at the time of the field work.

This analysis indicates that it is relevantly cheaper to conduct the camera-trapping method. The method allows time-saving in human resources as well. The overall cost saving, for the study's seven trials, was 17.5%. The margins allow the amortization of the camera traps after the fourth trial. After the amortization, the cost saving average is around 42.16% per trial.

4 Discussion

The use of camera-trapping in carcass persistence trials allows us to overcome some of the limitations associated with the conventional method. In the present study, there were not found significant differences between the correction factor values obtained with both methods.

In the study it was possible to obtain information about the removal agent 98 times in 156 removals, which is a similar result as observed in Paula et al. [26]. In only 3.85% of the carcasses was the known interval for removal bigger than 24 h. It was also possible to identify the removal agent in 90% of the times where the exact time of removal was obtained.

This kind of results allows those involved in the wind farm monitoring to better know the scavenger's populations present in their area of study and ultimately to open new questions about the carcass persistence and mortality assessment.

Obtaining uncensored data (exact time of removal) is only possible with the camera-trapping method, although the values obtained are not significantly different to those obtained with conventional carcass persistence trials. It should be looked closer, once the procedures in the camera setup are perfected, if this similarity is maintained.

This method also assures that the scavengers are not affected by the daily scent left by the technician, as might occur in the case of the conventional method, according to some authors [43, 44]. In this study we observed a few cases where the scavenger ignored the carcass right after its placement, which might be indicative of the human odor effect.

From a more ecological point of view, it should be mentioned that the great majority of the agents of removal from this study were nocturnal carnivores such as the fox (42.9%) and the beech marten (21.4%), which corresponds to the observed by Paula et al. [26]. This finding is not surprising since the fox is one of the most abundant carnivores in Portugal [45, 46], and its generalist and opportunistic diet [47, 48] makes the carcasses placed in the field an easy source of energy.

In the future it would be important to address the guild adjustment in case of big exogenous events in areas near the wind farms, like wild fires, and how it affects the carcass removal.

From the financial point of view, the quantitative comparison between the camera-trapping method and the conventional method, presented in Table 4, indicates that the camera-trapping method is significantly more advantageous economically. The main differences relate to the costs with field technicians, accommodation, and food.

It must be considered that the loss of data may occur, either by equipment malfunction or, as happened in a single occasion in this work, by equipment theft. When using camera traps, the trials must be planned with this risk under consideration. Sometimes memory cards can get full due to unwanted triggering, caused by the movement of vegetation due to wind [26, 31]. However, these problems can be solved or minimized by putting in place a series of good practices. We recommend placing the carcass at least 2 m away from the camera and avoiding the placement of the carcasses with objects on the back, to avoid burning the picture with the flash of infrared rays. The camera quality is also important to the success of the trial with this methodology, as well as the capacity to record in bigger memory cards. Cameras that have an option to take at least one photo per day, at predetermined hours, should also be used, to allow at least the same data one would have with the conventional methodology. Ultimately, when the methodology is improved to higher percentages, it is important to verify if there are still no differences found between the two methodologies.

This study lifts the possibility of considering removal agents in the calculation of correction factors and understanding how the ecological dynamics of each site or exogenous effects such as fire or human disturbance interaction may or may not influence withdrawal time.

5 Conclusion

The implications of this work are related mainly to the decision support on the methodology to be used in the calculation of the probability of carcass persistence. In this study, we show that the values obtained between the two methodologies are not significantly different for the study areas represented here, under the current camera performance obtained, and that in addition to the camera-trapping methodology, it is possible to enrich any study with the agents responsible for the removal and consequently to gain a better understanding of the ecosystem and the ecological dynamics that take place, even though this is not the current main focus of the removal tests and mortality calculations.

In the future this methodology may be standardized and advised in the guidelines of the different countries, as it happens already in other countries, like Scotland. The accuracy of the correction factor associated with the removal time will also be something to be looked closely into, so that the estimated mortality is increasingly close to the actual mortality. The fact that this methodology allows to obtain information about the agent of removal can still enrich the ecological knowledge of the study areas. The decision about which method to use is thus dependent on each case and each sampling site; however, this study may represent a tool to support and substantiate this decision.

References

1. Bernstein, M.A., Griffin, J., Lempert, R.: Impacts on Energy Expenditures of Use Technical report prepared for the energy future coalition. RAND Corporation, Santa Monica (2006)
2. Coelho, H., Mesquita, S., Mascarenhas, M.: How to design an adaptive management approach? In: Mascarenhas, M., Marques, A.T., Ramalho, R., Santos, D., Bernardino, J., Fonseca, C. (eds.) Biodiversity and Wind Farms in Portugal, pp. 208–224. Springer, Cham (2018)
3. Drewitt, A.L., Langston, R.H.W.: Assessing the impacts of wind farms on birds. Ibis. **148**, 29–42 (2006)
4. Arnett, E.B., Inkley, D.B., Johnson, D.H., Larkin, R.P., Manes, S., Manville, A.M., et al.: Impacts of Wind Energy Facilities on Wildlife and Wildlife Habitat. Bethesda, Maryland (2007a)
5. Arnett, E.B., Inkley, D.B., Johnson, D.H., Larkin, R.P., Manes, S., Manville, A.M., Mason, J.R., Morrison, M.L., Strickland, M.D., Thresher, R.: Impacts of Wind Energy Facilities on Wildlife and Wildlife Habitat Wildlife Society Technical Review 07-2. The Wildlife Society, Bethesda (2007b)
6. NRC: Environmental Impacts of Wind-Energy Projects. National Research Council of the National Academies. The National Academic Press, Washington, DC (2007)
7. Arnett, E.B., Baerwald, E.F., Mathews, F., Rodrigues, L., Rodríguez-Durán, A., Rydell, J., et al.: Impacts of wind energy development on bats: a global perspective (Chapter 11). In: Voigt, C.C., Kingstom, T. (eds.) Bats in the Anthropocene: Conservation of Bats in a Changing World. Springer, Berlin (2016)
8. Barclay, R.M.R., Baereals, E.F., Gruver, J.C.: Variation in bat and bird fatalities at wind energy facilities: assessing the effects of rotor size and tower height. Can. J. Zool. **85**, 381–387 (2007)

9. Kunz, T.H., Arnett, E.B., Erickson, W., Hoar, A.R., Johnson, G.D., Larkin, R.P., Strickland, M.D., Thresher, R.W., Tuttle, M.D.: Ecological impacts of wind energy development on bats: questions, research needs, and hypotheses. Front. Ecol. Environ. **5**, 315–324 (2007)
10. Baerwald, E.F., D'Amours, G.H., Klug, B.J., Barclay, R.M.R.: Barotrauma is a significant cause of bat mortalities at wind turbines. Curr. Biol. **18**, 695–696 (2008)
11. Santos, J., Marques, J., Neves, T., Marques, A.T., Ramalho, R., Mascarenhas, M.: Environmental impact assessment methods: an overview of the process for wind farm's different phases – from pre-construction to operation. In: Mascarenhas, M., Marques, A.T., Ramalho, R., Santos, D., Bernardino, J., Fonseca, C. (eds.) Biodiversity and Wind Farms in Portugal, pp. 35–86. Springer, Cham (2018)
12. Bernardino, J., Bispo, R., Costa, H., Mascarenhas, M.: Estimating bird and bat fatality at wind farms: A practical overview of estimators, their assumptions and limitations. N. Z. J. Zool. **40**(1), 63–74 (2013)
13. Erickson, W., Johnson, G., Young, D.A.: Summary and comparison of bird mortality from anthropogenic causes with an emphasis on collisions. Technical report, USDA Forest Service (2005)
14. Sinclair, K., DeGeorge, E.: Framework for testing the effectiveness of bat and eagle impact-reduction strategies at wind energy projects. National Renewable Energy Laboratory. Technical Report NREL/TP-5000-65624 (2016)
15. Smallwood, K.S.: Comparing bird and bat fatality-rate estimates among North American wind-energy projects. Wildl. Soc. Bull. **37**(1), 19–33 (2013)
16. Miller, A.: Patterns of avian and bat mortality at a utility scaled wind farm on the southern high plains. PhD thesis, Texas Tech University (2008)
17. Johnson, G.D., Erickson, W.P., Strickland, M.D., Sheperd, M.F., Sheperd, D.A., Sarappo, S.A.: Mortality of bats at a large-scale wind power development at Buffalo Ridge, Minnesota. The American Midland. Naturalist. **150**(2), 332–342 (2003)
18. Paula, J., Augusto, M., Neves, T., Bispo, R., Cardoso, P., Mascarenhas, M.: Comparing field methods used to determine bird and bat fatalities. In: Mascarenhas, M., Marques, A.T., Ramalho, R., Santos, D., Bernardino, J., Fonseca, C. (eds.) Biodiversity and Wind Farms in Portugal, pp. 135–149. Springer, Cham (2018)
19. Reyes, G.A., Rodriguez, M.J., Lindke, K.T., Ayres, K.L., Halterman, M.D., Boroski, B.B., et al.: Searcher efficiency and survey coverage affect precision of fatality estimates. J. Wildl. Manag. **80**(8), 1488–1496 (2016)
20. Huso, M.M.P., Dalthorp, D.: Accounting for unsearched areas in estimating wind turbine-caused fatality. J. Wildl. Manag. **78**(2), 347–358 (2014)
21. Korner-Nievergelt, F., Korner-Nievergelt, P., Behr, O., Niermann, I., Brinkmann, R., Hellriegel, B.: A new method to determine bird and bat fatality at wind energy turbines from carcass searches. Wildl. Biol. **17**(4), 350–363 (2011)
22. Morrison, M.: Searcher Bias and Scavenging Rates in Bird/Wind Energy Studies Technical Report NREL/SR-500–30876, p. 5. National Renewable Energy Lab, Golden Colorado (2002)
23. Smallwood, K.S.: Estimating wind turbine-caused bird mortality. J. Wildl. Manag. **71**(8), 2781–2791 (2007)
24. Travassos, P., Costa, H.M., Saraiva, T., Tomé, R., Armelin, M., Ramírez, F.I., et al.: A energia eólica e a conservação da avifauna em Portugal. SPEA, Lisboa (2005)
25. Bispo, R., Bernardino, J., Marques, T.A., Pestana, D.: Discrimination between parametric survival models for removal times of bird carcasses in scavenger removal trials at wind turbines sites. In: Lita da Silva, J., et al. (eds.) Advances in Regression, Survival Analysis, Extreme Values, Markov Processes and Other Statistical Applications, Studies in Theoretical and Applied Statistics. Springer, Berlin (2013)
26. Paula, J.J., Bispo, R.M., Leite, A.H., Pereira, P.G., Costa, H.M., Fonseca, C.M., Bernardino, J.L.: Camera-trapping as a methodology to assess the persistence of wildlife carcasses resulting from collisions with human-made structures. Wildl. Res. **41**(8), 717–725 (2015)
27. Silva, B., Barreiro, S., Alves, P.: Factors influencing carcass removal at wind-farms, in mountain areas of the center-north region of Portugal. In: Abstracts XIst European Bat Research Symposium, pp. 18–22. Agosto, Roménia (2008)

28. APA: Guia para a Avaliação de Impactes Ambientais de Parques Eólicos (2009)
29. Bernardino, J., Bispo, R., Torres, P., Rebelo, R., Mascarenhas, M., Costa, H.: Enhancing carcass removal trials at three wind energy facilities in Portugal. Wildl. Biol. Pract. **7**, 1–14 (2011)
30. Jones, A.: Animal scavengers as agents of decomposition: the postmortem succession of Louisiana wildlife. Masters thesis, Louisiana State University, Baton Rouge, LA, USA (2011)
31. Smallwood, K.S., Bell, D.A., Snyder, S.A., DiDonato, J.E.: Novel scavenger removal trials increase wind turbine–caused avian fatality estimates. J. Wildl. Manag. **74**(5), 1089–1097 (2010)
32. DeVault, T.L., Brisbin Jr., I.L., Rhodes Jr., O.E.: Factors influencing the acquisition of rodent carrion by vertebrate scavengers and decomposers. Can. J. Zool. **82**, 502–509 (2004)
33. Cunningham, P.D., Brown, L.J., Harwood, A.J.: Predation and scavenging of salmon carcasses along spawning streams in the Scottish Highlands. Final report for the Atlantic Salmon Trust (2002)
34. Kostecke, R.M., Linz, G.M., Bleier, W.J.: Survival of avian carcasses and photographic evidence of predators and scavengers. USDA National Wildlife Research Center – Staff Publications. Paper 510 (2001)
35. SNH: Monitoring the Impact of Onshore Wind Farms on Birds Guidance Note. Scottish Natural Heritage, Scotland (2009)
36. Duffy, K., Steward, M.: Turbine Search Methods and Carcass Removal Trials at the Braes of Doune Windfarm Natural Research Information Note 4. Natural Research Ltd., Banchory (2008)
37. Cox, D.R., Oakes, D.: Analysis of Survival Data, vol. 21. CRC Press, Boca Raton (1984)
38. Bispo, R., Palminha, G., Bernardino, J., Marques, T., Pestana, D.: A new statistical method and a web-based application for the evaluation of the scavenging removal correction factor. In: VIII Wind Wildlife Research Meeting on Proceedings, pp. 19–21. Lakewood. October (2010)
39. Burnham, K.P., Anderson, D.R.: Multimodel inference: understanding AIC and BIC in model selection. Sociol. Methods Res. **33**(2), 261–304 (2004)
40. Huso, M.: An estimator of wildlife fatality from observed carcasses. Environmetrics. **10**(22), 318–329 (2010)
41. R Development Core Team R Development Core Team: R: A Language and Environment for Statistical Computing. R Foundation for Statistical Computing, Vienna (2013)
42. Bates, D., Maechler, M., Bolker, B., Walker, S.: Fitting linear mixed-effects models using lme4. J. Stat. Softw. **67**, 1–48 (2015)
43. Linz, G.M., Bergman, D.L., Bleier, W.J.: Estimating survival of song bird carcasses in crops and woodlots, p. 842. USDA National Wildlife Research Center-Staff Publications (1997)
44. Ponce, C., Alonso, J.C., Argandoña, G., García Fernández, A., Carrasco, M.: Carcass removal by scavengers and search accuracy affect bird mortality estimates at power lines. Anim. Conserv. **13**(6), 603–612 (2010)
45. Blanco, J.C.: Mamíferos De España, vol. I and II. Ed. Planeta, Barcelona (1998)
46. Palomo, L.J., Gisbert, J., Blanco, J.C.: Atlas y Libro Rojo de los Mamíferos Terrestres de España. Dirección General de Conservación de la Naturaleza – SECEM-SECEMU, Madrid (2007)
47. Dell'Arte, G.L., Laaksonen, T., Norrdahl, K., Korpimaki, E.: Variation in the diet composition of a generalist predator, the red fox, in relation to season and density of main prey. Acta Oecol. **31**, 276–281 (2007)
48. Webbon, C.C., Baker, P.J., Cole, N.C., Harris, S.: Macroscopic prey remains in the winter diet of foxes *Vulpes vulpes* in rural Britain. Mammal Rev. **36**, 85–97 (2006)

Lost in Bias? Multifaceted Discourses Framing the Communication of Wind and Wildlife Research Results: The PROGRESS Case

Jessica Weber, Juliane Biehl, and Johann Köppel

Abstract In times of increasingly selective interpretations of research results, sound decision-making in environmental management can be complicated. Actors compete for opinion leadership over relevant findings. A recent example is the agitated debate over the study "Prognosis and assessment of collision risks of birds at wind turbines in northern Germany" (PROGRESS). This debate, arising in the science-policy-practice interface, challenges research results and subsequent implications for planning and decision-making processes. We used the PROGRESS discourse to conduct a "frame analysis" and identified relevant actors, their argumentation patterns and potential motives. Discourses reveal patterns of perception that can be spotted by content analyses of available statements. It became obvious that research results do not necessarily just reduce knowledge gaps. The frames "wildlife protection and conservation" and "research methods" were of major importance in the discourse, addressing the paramount question of whether wind energy development actually affects raptor populations. The discourse indicated, *inter alia*, that uncertainties still revolve around collision risk estimators, demographic population models, the efficacy of mitigation measures and whether the challenges at hand might be dealt with on a case-by-case or a metalevel planning approach. In conclusion, the question remains: Who might unanimously communicate both findings and implications? A thorough and bias-sensitive communication of research results might foster sound decision-making processes in the wind and wildlife sector, particularly based on "best available science" efforts.

Keywords Science-policy-practice interface · Frame analysis · Wind energy and wildlife · Scientific communication · Sound decision-making · Interpretation of research results · Best available science

J. Weber (✉) · J. Biehl · J. Köppel
Environmental Assessment and Planning Research Group, Berlin Institute of Technology (TU Berlin), Berlin, Germany
e-mail: j.weber@campus.tu-berlin.de; juliane.biehl@tu-berlin.de; johann.koeppel@tu-berlin.de

R. Bispo et al. (eds.), *Wind Energy and Wildlife Impacts*,
https://doi.org/10.1007/978-3-030-05520-2_12

1 Introduction

Lost in bias? This study uses a recent debate in Germany over the effects of wind energy on the common buzzard (*Buteo buteo*) and red kite (*Milvus milvus*) to raise the awareness of adequate communication and how decision-making processes may be influenced by actors' diverging interpretation patterns. It is obvious that an opinion battle over the interpretation of scientific research results comes not without risk [1, p., 261], as "[...] there is no shortage of politicized research topics, where the motives of researchers and the interpretation of their findings are fiercely disputed". In the wind energy and wildlife arena, research results play a pivotal role in overcoming the science-policy-practice gap [2, 3 for conservation]. They are indispensable in decision-making processes on the siting of wind turbines, management actions and for the identification of further research requirements. The way in which relevant information is interpreted can substantially affect planning practice and attitudes [4, 5].

However, a range of factors can trigger bias in the interpretation of information [1]. For example, scientific information can be interpreted due to individual bias in reporting over- or underestimations of risks [6]. Moreover, publication bias can occur when findings do not seem to be compatible with the state of knowledge or are not favourable and are retained [7]. In addition, individuals with a range of professional backgrounds and education generally might value scientific information differently [1, 8]. Due to the relevant mind-set of respective actors, important facets of any given information might be considered in divergent ways. The latter was well demonstrated, for example, in a discourse analysis on perceptions of maize cropping [8].

1.1 Study Motivation

Opinion leadership has also become particularly relevant in wind energy- and wildlife-related decision-making processes. As soon as an array of different actors is involved, a wide variety of interpretation patterns emerges [1]. Usually various stakeholder groups (e.g. planners and developers, agencies, consultants, nongovernmental organisations, civil opponent groups, other experts) take part in the approval processes for wind turbines, thus yielding manifold interpretation patterns.

Each actor considers information in a certain light – conservationists may tend to be more critical towards possible impacts, while project developers may be concerned with constraints for wind energy development [9]. While this is not a problem in itself, problems evolve when actors claim leadership in interpretation. Expected or desirable results should not be preferentially reported, as publishers of research results are also required to refrain from withholding information [6]. At the end of the day, ensuring "good" management decisions becomes complicated by the sheer competition on how to understand and properly work with research results.

Consequently, we shifted the focus on the dynamic tension inherent in predetermined views and conducted an initial discourse analysis concerning wind energy and wildlife – based on Linhart and Dhungel's [8] methodological approach. According to Linhart and Dhungel [8], a discourse analysis investigates the interests of society: How actors communicate in public is especially relevant as the latter can act as a forum to gain interpretive standing. At the same time, public opinion can be influenced to further individual interests. It is precisely at this point that a discourse analysis can help to analyse the argumentation patterns, relevant predominance and motives as well as implications [8].

In consequence, there are multiple reasons to raise the awareness of the communication and interpretation of information with a discourse analysis. The adage "The one who shouts the loudest is not necessarily right, but is the most likely to be heard" must be considered as well. A knowledge-based society is placed in jeopardy when the importance of unbiased information becomes challenged [10]. Many of the questions that arise in the science-society interface set out to understand how influential any interpretation can be and to analyse wind energy and wildlife implications.

1.2 The "PROGRESS" Case Study

A recent example for the multifaceted interpretation of scientific research results is the agitated debate in 2016 that took place in Germany over comprehensive research on the "Prognosis and assessment of collision risks of birds at wind turbines in northern Germany" (short: PROGRESS) [11–13]. One newspaper headline reads prior to the publication of the PROGRESS study: "The study is not published at all; however, the excitement is already big" [14].

The underlying reasons were that PROGRESS sheds new light on the extent and consequences of bird mortality at wind turbines [13], after 5 years of research (1 November 2011 to 30 June 2015) [11, 12]. The study conducted an extensive *fatality search* in the northern German lowland, which is presently the hub of wind energy development in Germany [15]. Moreover, *population models* were developed to quantify population effects, building on fatality estimators. PROGRESS sheds specific light on the common buzzard (*Buteo buteo*) – a raptor species, which had not been considered substantially as wind energy sensitive before – as the study estimates a population decline for both the endangered red kite (*Milvus milvus*) and the common buzzard [11–13] (Fig. 1).

In the ensuing discussion in public, the question arose as to how far new findings should ultimately be applied in wind farm approval practice [16]. Required by jurisdiction, scientific findings have to be taken into account in decision-making processes. The European Court of Justice (ECJ) decided to consider "the best scientific knowledge" in decision-making processes with regard to the EU Habitats

Fig. 1 The common buzzard (*Buteo buteo*), left, and the red kite (*Milvus milvus*), right, soaring. (Copyright Jessica Weber, 2016)

Directive (92/43/EEC).[1] A similar decision by the German Federal Administrative Court (BVerwG) requires to use "exclusively scientific criteria" in decision-making processes for species protection[2] [17].

It is apparent that both research results and how they are understood and interpreted are important in the science-policy-practice interface for wind energy and conservation. After all, regardless of the content of research, how actors in society perceive and handle information can dominate management practices and approval processes. Despite this, the risks are likely to remain underestimated so far.

1.3 Objectives

In order to identify potential bias in the communication of wind energy and wildlife research and to highlight possible implications, we selected the discourse surrounding the "PROGRESS" case study. We aimed at identifying specific mind-sets of actors and anticipated that topics, which various actors consider relevant within the discourse, can be categorised into so-called frames [8]. Moreover, an illustrative typology of potential motives helped to address the questions:

1. Which actors participated in the relevant discourse framing wind energy and wildlife research? What were the actors' motives?
2. What challenges arise from the discourse for scientists, decision-makers, practitioners and the general public?
3. What actions could be taken to overcome such challenges in the science-policy-practice interface?

[1]CJEU, judgement of 7 September 2004, C 127-02, paragr. 54, 61.

[2]BVerwG, judgement of 9 July 2008, 9 A 14.07, paragr. 64.

2 Methods

2.1 Frame Analysis: Indicating Actors and Subjects in the Discourse

According to Gee [18, p. 18] "[. . .] [a] discourse analysis can illuminate problems and controversies in the world [. . .]" as the language is interpreted as key way humans deal with social goods. However, there are several methodological approaches [18–20]. While some only analyse the content of the language used, i.e. the topics discussed by actors or the media, other approaches focus more on the grammar of language and its implications [18]. For example, Gamson and Modigliani [21, p. 3] underline that a "[. . .] media discourse can be conceived as a set of interpretive packages that give meaning to an issue. A package has an internal structure. At its core is a central organizing idea, or frame, for making sense of relevant events, suggesting what is at issue".

In this study we conducted a frame analysis evaluating frequently discussed topics surrounding the PROGRESS case study [8]. The method was applied by Hess et al. [22] as well as by Linhart and Dhungel [8] and had been described initially by Gamson and Modigliani [21]. It assumes that discourses comprise different patterns of perception and interpretation, which can be categorised by content (e.g. in our case, conservation, economy, policy and jurisdiction). Bias can result from different actors with different perspectives participating in a discourse. Every "frame" encloses a decisive subject that various actors consider important within the discourse, representing a specific mind-set of an actor and shifting the focus on specific argumentation patterns [8]. "In other words, framing can be seen as the process of reducing and simplifying complex realities to enable action" [23, p. 21 referring to 24].

We analysed publicly available press releases and articles about PROGRESS, thereby identifying involved actors and collecting arguments submitted in the discourse. Regardless of the standpoint of the actors, identified arguments were ultimately grouped in a frame, and actors were anonymised. Moreover, the identified actors submitting arguments were grouped according to their professions in order to illustrate likely motives of actors with similar perspectives.

Within each frame, argumentation patterns can vary standpoint-wise. For example, actors can interpret research results negatively as well as positively. One actor may believe that mitigation measures are sufficient, while another actor feels that the enhancement of mitigation measures is a priority. This means that different argumentation patterns within a superior mind-set grouping of various actors (i.e. frames) can also occur. Argumentation patterns can be controversial due to positive or negative directions of verbal statements (+/−). With regard to the PROGRESS case study, this is mainly a result of actors' opinions about the two major methodological approaches (fatality searches and population models in PROGRESS).

Dividing perspectives and argumentation patterns into a frame can be conducted by either rough or precise classification [8]. For this analysis, we identified a rough structure of frames, which represent the most frequently discussed topics. In this way, the most commonly featured mind-sets and argumentation patterns of actors were juxtaposed. The rough division of mind-sets into frames enables to analyse these contradicting mind-sets and to identify patterns of argumentation and individual interpretations of research results.

2.2 Typology: Indicating Argumentation Patterns and Motives

The question arose how contradicting interpretations of PROGRESS' research results were reasoned in each frame. In order to probe the argumentation patterns and motives of interpretation leadership in the PROGRESS case, this was assessed by the voiced arguments [20]. According to the frame analysis methodology by Linhart and Dhungel [8], firstly, we compiled how every actor brought forward distinctive argumentation. Secondly, the underlying subtext of these patterns was analysed and classified to capture potential motives of participating in the public discourse. Both methodological approaches, the "framing" and "typology" of subjects and argumentation patterns, were combined analytically in order to emphasise the tensions and uncertainties in the wind energy and wildlife arena resulting from divergent perceptions within the frames (i.e. subjects) – for example, whether the contemporary approval practice should apply the knowledge of the PROGRESS research project or not.

The typology was adapted in line with studies mainly in the field of psychology that categorise different social behaviours of people [25–27]. Four categories eventually evolved to group and visualise likely motives and beliefs of actors by analysing the stakeholders' contents and wording of voiced arguments:

- "Preservers", who consider the current permission policy to be adequate.
- "Change agents", who point out adjustment potential in German permitting practice due to the results of PROGRESS' population models.
- "Objectifiers", who opine that the modelled collision risks have only stochastic character – the PROGRESS study lacks topicality.
- "Relativists", who consider other mortality factors more relevant than wind energy (e.g. agriculture, transmission grid).

The category "preservers" comprises stakeholders, who generally chose positive words towards PROGRESS' results of the *fatality searches* (i.e. a general acceptance of fatality findings below wind turbines), but chose increasingly negative wording regarding PROGRESS' *population modelling*. In contrast, actors categorised under "change agents" focused their argumentation on the possible population decline identified by PROGRESS' *population modelling*. "Change agents" therefore voiced interest in adjusting the current planning practice. The "objectifiers" resemble the preservers' category in wording and argumentation, since both generally accept the results of PROGRESS; however, “objectifiers” negate the *population modelling* results. Besides, “objectifiers” interpret the research results to support the current state of knowledge. Moreover, “relativists” perceive that other land uses would have higher mortality rates and impacts on the species’ population levels than wind energy (e.g. transmission lines, agriculture).

Again, different motives can be ascribed to actors reasoning the voiced arguments. For example, some actors advocate for changes in the current approval practice, while others reject relevant modifications. In addition, we analysed how well arguments were supported in the discourse, for instance, with regard to stated sources, supporting or refuting statements.

2.3 Data Basis and Limitations

A prepublication of PROGRESS’ research results, in the reputable newspaper “Süddeutsche Zeitung” in early 2016 [28], indicates a starting point to the debate. Eventually, the complete report on PROGRESS’ research results was published in June 2016 [11–13]. Our analysis includes published data within a one-year period from January 2016 to January 2017.

Around 25 press articles and statements in magazines, including daily and weekly newspapers, were examined, as they are known to mirror the perception of their audience and are intensely observed by political decision-makers [8]. Furthermore, the communication of panellists in a discussion forum for interested stakeholders in the wind energy and conservation sector was analysed. The forum was organised by “Onshore Wind Energy Agency” and was held on 17 November 2016 [16].

Voiced statements were mostly gathered with a keyword search on the Internet. Generalised keywords included “PROGRESS Rotmilan” (red kite), “PROGRESS Mäusebussard” (common buzzard) and “PROGRESS Stellungnahmen” (statements). Evaluating all available data was beyond the scope of feasible activities. Hence, the analysis is based on a sample set of the available data, which picks out central arguments. The analysis does not claim to be complete; instead it intends to provide an overview of the German discourse within the relevant science-policy-practice interface.

However, the choice of information can influence the actors identified as part of the “PROGRESS” discourse. Thus, an all-encompassing analysis of all involved actors and all voiced arguments within the discourse was beyond the scope of feasible activities. Hence, the number of identified actors and arguments cannot be evaluated in terms of their representativeness and importance. For example, it would be valuable to analyse the weight and standpoint of actors and their arguments in a discourse.

It should also be noted that the sheer number of arguments is not an indicator of relevance in a discourse. The fact that a single argument is voiced and accordingly categorised in a frame should not diminish its importance, even if its motive cannot be further quantified (cf. frame “climate protection”, Sect. 3.2). The diversity of proposed arguments may also be an indication of remaining uncertainties in the wind energy and wildlife sector towards the identified frames.

3 Frame Analysis and Typology: Indicating Actors and Subjects in the Discourse and Typology of Motives

3.1 Indicating Actors

Table 1 provides an overview of the identified and anonymised actors and illustrates that the PROGRESS discourse is dominated by five groups of actors: (1) policy makers and agencies, (2) associations, (3) experts (environmental consultants and scientists), (4) project developers and (5) others.

All in all, actors presented in Table 1 can be labelled as *key actors*, as it can be assumed that the discourse surrounding PROGRESS was not only carried out in public but also in private [8]. Whom the media cited is influencing the perception of the public concerned to spread out information about the PROGRESS case and its new findings for the interested public [8, 29]. Thus, the actors concerned are influenced by media coverage, in the course of the dissemination of PROGRESS' information to the general public. This also means that information about who is participating in the discourse might have been presented selectively by the

Table 1 Identified actors in the PROGRESS discourse and typology of their professional background

Policy makers and agencies	Associations and foundations	Experts (environmental consultants and scientists)	Project developers	Others
Ornithological Station	Wind Energy Association	Landscape architecture consultancy	Project development consultancy	Journalists
Former member of the parliament (Green Party)	Association for Bird Protection	Environmental planning consultancy	Project development and environmental consultancy	Independent network of scientists and parliamentarians
Lower nature protection agency	Nature Conservation Association	Onshore Wind Energy Agency		
Federal Agency for Nature Conservation	Society for the Conservation of Owls			
A ministry	Wildlife Foundation			

media as well. For example, in our analysis, the two groups "policy makers and agencies" and "associations and foundations" were the largest participating groups in the discourse. They were possibly seen and heard more often than, for example, scientists. These actors seemed to dominate the discourse, framing the perception of the research results.

The research consortium of the PROGRESS study consisted of three consultant groups and one professor of animal behaviour at a university [11, 12]. The research consortium was not listed in Table 1, since it took a key position in the discourse anyway. It is important to note that the discourse was influenced as well by a statement-response pattern between the research consortium and the actors indicated in Table 1. Hence, the identification of argumentation patterns and the framing of arguments also encompassed the communication of the research consortium.

For example, on the discussants' side, one actor stood out due to a strong presence [30, 31]. This actor acted as an "opponent" and denied PROGRESS' findings, especially with regard to the estimated decline in red kite population. Reference was made to a contrasting document by the same actor [31, p. 3], which states that "[...] a possible threat to the continued existence of the red kite cannot be justified from the state of knowledge". On the contrary, this actor stated to have observed a high compatibility between the red kite and wind energy facilities [31].

Generally speaking, these conflicting argumentation patterns between the findings of the research consortium and other perceptions could be traced through the PROGRESS discourse as a theme. For example, a former member of the parliament supported the argumentation pattern of the aforementioned opponent to the PROGRESS study [32].

3.2 Indicating Argumentation Patterns Within Identified Frames and Typology of Motives

The argumentation patterns of the identified actors were grouped into the following five frames, which indicate the five most discussed topics. A frame represents a package of interpretative subjects in which divergent argumentation patterns are able to contrast with one another [21]. The identified frames are:

- Wildlife protection and conservation
- Research methods used in PROGRESS
- Economy
- Policy makers and jurisdiction (responsibilities)
- Climate protection

Actors' argumentation patterns were categorised in the frame "wildlife protection and conservation", which contained subjects such as the interactions between wind energy and wildlife (e.g. collisions), issues related to environmental planning (e.g. mitigation measures, scale of planning, species relevant to planning) and the legal background of conservation (e.g. related to the Federal Nature Conservation Act) – since legislation shapes the scope of action for wildlife protection and environmental planning (e.g. executive leeway, population level vs. case-by-case level). The frame "research methods used in PROGRESS", in contrast, summarised argumentation patterns that concerned methodological approaches of the PROGRESS case, such as population modelling and fatality searches.

Moreover, argumentation patterns relating to economic interests were grouped within the frame "economy", while the frame "policy and jurisdiction (responsibilities)" dealt with the question, which political actors were responsible for bringing clarity and certainty in the discourse around PROGRESS' research results. This frame thus should not be confused with legislative subjects relating to conservation in the frame "wildlife protection and conservation". Lastly, actors' argumentation patterns, which dealt with the discrepancy between wildlife and climate protection, were categorised in the frame "climate protection".

Generally the argumentation patterns can be classified to support or deny PROGRESS' research results (+/−).

3.2.1 Frame "Wildlife Protection and Conservation"

In Table 2, the actors' main argumentation patterns are shown for the frame "wildlife protection and conservation". It became apparent that subjects mainly enclosed the legislation of the protection of endangered species and planning-related consequences for project developments. This is due to the fact that its scope is determined by legislative prohibitions, e.g. by the Federal Nature Conservation Act. The frame was categorised with regard to two partial results of the PROGRESS study: on the one hand, the population models for the red kite and common buzzard and, on the other hand, the results of the fatality searches and collision risk estimates. The direction in which actors communicated is indicated in Table 2.

It is becoming apparent that the frame encompasses contradicting points, reflecting the question of the supportability of PROGRESS' research results. It is notable that some actors took on the role of opponents to the PROGRESS' research results (e.g. individual actors in the group "project developers" and "policy makers and agencies"), while actors involved in nature conservation associations took an active part in this discourse by defending PROGRESS' research results.

Change Agents

By accepting the research results, including the fatality searches and population models, some actors perceived the common buzzard to be threatened by wind energy facilities (Nature Conservation Association [41]; Lower Nature Protection Agency [16]). Here, a group of actors was identified as "change agents". As the common buzzard is not considered a relevant species in wind energy planning and zoning yet, the discussion revolved around legal security and the question whether a

Table 2 Argumentation patterns within the frame "wildlife protection and conservation" to support or reject PROGRESS' research results

Frame "wildlife protection and conservation"	
+ (research results are valid, mainly referring to population models)	− (research results are not valid or controversial, mainly referring to fatality searches)
PROGRESS sheds a new light on the interactions between wind energy and raptors and possible threats. Therefore, the common buzzard should be considered in planning as a species of relevance (Lower Nature Protection Agency [33])	Since fatalities occur, impacts of wind energy seem to be confirmed; nevertheless, these are only stochastic events, which cannot be forecasted and prevented through planning (Ornithological Station [35]; Environmental planning consultancy [36])
This knowledge should be applied in planning practice (Landscape architecture consultancy [16]; Nature Conservation Association [34])	
Due to possible population declines of the red kite and common buzzard, the current planning practice does not seem to be sufficient (Nature Conservation Association [34]; Lower Nature Protection Agency [37]; A ministry 2017 [38]; Society for the Conservation of Owls [39])	The current approval practice is successful since there are only low fatality findings below wind turbines (Independent network of scientists and parliamentarians [32]; Project development and environmental consultancy [30]; Former member of the parliament (Green Party) [40])
	The predictions of negative population developments for the red kite and common buzzard are false (Project development and environmental consultancy [30], Former member of the parliament (Green Party) [40]; Independent network of scientists and parliamentarians [40]; Society for the Conservation of Owls [39])
Higher planning levels should be taken into consideration for environmental planning (Association for Bird Protection [37]; Nature Conservation Association [41])	Supra-regional scales for planning are possibly ineffective when considering population consequences (Federal Conservation Agency [16])
Whether the Federal Act for Nature Conservation should open up to consider population effects rather than effects on individuals should be discussed (A ministry [38])	
It is possible that the distance recommendations for nesting and breeding areas for bird species are not sufficient (Nature Conservation Association [41]; Wildlife Foundation [42])	Distance recommendations for nesting and breeding areas for bird species should not be updated with regard to PROGRESS research results (Wind Energy Association [37]; Federal Conservation Agency [33]; A ministry [38]; Association for Bird Protection [16])

stronger protection status would be required. Demands to implement PROGRESS' research results in planning practices were being raised. The fear of a potential future population decline was reflected as well, with petitioning in support of the precautionary principle in environmental law and therefore for a consideration of the new research findings in planning practice. Moreover, some actors perceived the research results as an indicator of insufficient approval practice with regard to mitigation measures (Society for the Conservation of Owls [39]). In this case, actors might have referred to results of the study's population modelling.

Relativists

In contrast, other actors rejected the results of the population models (Independent network of scientists and parliamentarians [32]; Former member of the parliament (Green Party) [40], Project development and environmental consultancy [30]; Federal Conservation Agency [33]; Project development consultancy [43]). This was expressed, for instance, by referring to contrary results of other studies and the "current state of knowledge" (Project development and environmental consultancy [30, 31]). However, sources that supported the actors' statements were not peer-reviewed [31] or were not widely cited.

Furthermore, other actors shared the impression that the impacts of wind energy were secondary when compared with other risk factors, such as:

- Agriculture (Nature Conservation Association [34], Project development and environmental consultancy [30])
- Grid development (Project development and environmental consultancy [30])
- The positive effects of mitigation measures for birds (Project development and environmental consultancy [30], Environmental planning consultancy [36])

We thus identified the group of "relativists". By analysing the group of actors communicating this perception of the research results, it became obvious that some actors may have tried to prevent likely restrictions to the development of wind energy, which could have evolved from the PROGRESS discourse.

In contrast, other arguments highlighted the actors' own methodological approaches to estimate the sensitivity of the common buzzard to wind energy, which yield only a medium mortality-hazard-index (Federal Agency for Nature Conservation [33]).

Preservers

As PROGRESS' fatality searches with low findings were perceived to confirm the approval practice, a call for the "status quo" was made. For example, an environmental planning consultancy perceived PROGRESS' collision estimates as stochastic events [36]. This research result is conceived to concur with the current state of knowledge with regard to prior studies [44–46]. PROGRESS' population models were not considered to meet methodological standards. Therefore, these actors seem to act as "preservers".

Objectifiers

In fact, the low fatality counts of PROGRESS below wind turbines were also perceived positively by actors (Wind Energy Association [34, 47]). Generally, the study was supported in this regard, as its results objectify the debate between wind energy opponents and project developers with regard to species protection (Wind Energy Association [48]). This means that the state of knowledge regarding the interaction between wind energy and wildlife seemed to be verified and exculpatory evidence was brought forward.

3.2.2 Frame "Research Methods Used in PROGRESS"

The analysis of argumentation patterns showed once again that actors' standpoints were conflicting and were classified into two distinguishable thematic blocks. On the one hand, the study's methodology was accepted; on the other hand, actors questioned the investigation methodology and also the study's results (Table 3).

Table 3 Argumentation patterns within the frame "research methods used in PROGRESS" to support or reject PROGRESS' research results

Frame "research methods used in PROGRESS"	
+ (research results are valid, mainly referring to population models)	– (research results are not valid or controversial, mainly referring to fatality searches)
PROGRESS is the first sound and comprehensive analysis with regard to the interactions between birds and wind energy (Wind Energy Association [48]; Nature Conservation Association [41])	Suitable, exact investigation methods have not been applied (Project development and environmental consultancy [30]; Former member of the parliament (Green Party) [40], Ornithological Station [35])
PROGRESS delivers a respectable database which makes clear that the approval practice fulfils the demands of the German Federal Nature Conservation Act (Wind Energy Association [47])	The investigation methodology is afflicted with uncertainties and weaknesses (Federal Nature Conservation Agency in [33]; Ornithological Station [35]; Agency for onshore wind energy [43])
There is no essential new result; yet, interesting observations and a huge methodological progress have been made (Wind Energy Association [37])	PROGRESS is afflicted with incorrect academic judgements (arithmetic mistakes), and there are content-related contradictions (Project development and environmental consultancy [30]; Independent network of scientists and parliamentarians [32]; Ornithological Station in [35]; Society for the Conservation of Owls [39])
There are also other studies, which are based on unevaluated models (Environmental planning consultancy [36])	Other studies have partly a higher control intensity (Ornithological Station [35])

Relativists

Based on the statement that PROGRESS had not applied suitable investigation methods – since the methods used were not accurate enough (e.g. in comparison to telemetry data, Project development and environmental consultancy [30]; Former member of the parliament (Green Party) [40], Ornithological Station [35]) – the possible threat to the red kite and common buzzard was not accepted to be true. Reasons for this were seen, for instance, in the unsuitable data used as basis for the extrapolation of fatalities for future population changes. Actors pled for more accurate research methods, such as the telemetry of raptors (Project development and environmental consultancy [30]). Moreover, reasoning was based on the poor scientific evaluation of research results, such as arithmetic mistakes (Project development and environmental consultancy [30]; Independent network of scientists and parliamentarians [32]; Ornithological Station [35]; Society for the Conservation of Owls [39]).

Moreover, one actor also stood out for critically reflecting the methodological approach of PROGRESS in suggesting that the research consortium had inadequately presented the research results by, for example, relativising fatalities (Society for the Conservation of Owls [39]).

Objectifiers

In contrast, actors in the cluster "objectifiers" generally neglected the research results of the population models, but voiced belief in a sound and comprehensive analysis of the fatality searches (Wind Energy Association [47]). Since PROGRESS found fewer fatalities than expected, actors took a positive attitude regarding the compatibility of avifauna and wind turbines. The sensitivity of the PROGRESS case study was minimised, and to some degree, research results were selectively cited. Some actors therefore tried to objectify the discourse around PROGRESS.

However, PROGRESS was simultaneously perceived as an advanced approach in methodology (Wind Energy Association [47]). For example, actors underlined the study's comprehensiveness (Nature Conservation Association [41]). Still, it may be assumed that different actors referred also to different methodological approaches (e.g. fatality searches or population modelling), according to the respective outcome. In any case, some actors diverted attention to comparable studies, which had a higher (methodological) control intensity, while other actors discussed the utility and security of models (Ornithological Station [35]; Environmental planning consultancy [36]).

Preservers

The argumentation patterns of "preservers" supported the debate on the applicability of research results for the current approval practice and Federal Nature Conservation

Act. Consequently, the Ornithological Station referred once more to the distance recommendations for nesting and breeding areas for bird species [49] to avoid planning uncertainties (Ornithological Station [35]). These recommendations were based on a literature review on the impacts of wind energy and birdlife [49]. Thus, argumentation patterns seemed to underline and engage in preserving the safety of current planning practice, arguing that the recommendations for nesting and breeding bird species already considered the state of knowledge.

3.2.3 Frame "Economy"

The argumentation patterns relating to economic interests are presented in Table 4. We identified a contrary pattern of perception, as on the one hand, results were designated as economy diminishing for the wind energy sector, while on the other hand actors claimed that the study lacked a critical eye regarding the protection of wildlife. Anyhow, one actor alleged the PROGRESS research consortium's economy friendliness.

Relativists

The argumentation patterns show that there seemed to be a lack of trust in existing structures and in policy makers. Additionally a lack in objectivity of the research project was mentioned, since PROGRESS was funded by federal agencies

Table 4 Argumentation patterns within the frame "economy" to support or reject PROGRESS' research results

Frame "economy"	
+ (research results are valid, mainly referring to population models)	– (research results are not valid or sensitive, mainly referring to fatality searches)
The research consortium is perceived to be independent and professional (Nature Conservation Association [41])	There is a missing objectivity of the research consortium; the PROGRESS study is business-friendly as it relativises its findings and the distance recommendations for nesting and breeding areas for bird species (Society for the Conservation of Owls [39])
	The research consortium is perceived to follow financial interests and to communicate an excessive emotional connection (to birdlife) (Project development and environmental consultancy [30])
	The implications of the research results are feared since the consideration of the common buzzard as a species relevant for planning may slow down future wind energy development (Journalists [50])

[11, 12]. Based on this perception, some actors relativised the findings of the PROGRESS study from the start. The research consortium was criticised for its structure and methodological approach (Project development and environmental consultancy [30]). For example, PROGRESS' findings were regarded to be too friendly regarding the wind energy sector (Society for the Conservation of Owls [39]).

In contrast, other actors feared the implementation of PROGRESS – being a new scientific finding – into planning practice, which might imply economic restraints for wind energy (Journalists [50]). Furthermore, other actors perceived PROGRESS' findings to communicate an excessive emotional connection to birdlife (Project development and environmental consultancy [30]). Therefore, some actors seemed particularly vigilant in the discourse.

3.2.4 Frame "Policy Makers and Jurisdiction (Responsibilities)"

The frame "policy makers and jurisdiction (responsibilities)" dealt with the question of who is responsible for bringing clarity and certainty in the discourse around PROGRESS' research results. Thus, planning levels and political actors are addressed in Table 5.

Change Agents

In the discourse, actors addressed the responsibility of legislation to interpret and to deal with PROGRESS' research results to ascertain security in planning practice (research consortium [51]; Nature Conservation Association [33]). In particular, some actors hinted at a required differentiation between science (in a sense of neutral, analytical work) and the normative interpretation of research findings (e.g.

Table 5 Argumentation patterns within the frame "policy makers and jurisdiction (responsibilities)" to support or reject PROGRESS' research results

Frame "policy makers and jurisdiction (responsibilities)"	
+ (research results are valid, mainly referring to population models)	– (research results are not valid or sensitive, mainly referring to fatality searches)
Actors perceive that the legislation at federal state level is in place to decide about whether and how PROGRESS' findings are incorporated in planning practice (Federal Agency for Nature Conservation [33])	–
It is perceived that judges feel insecure when ruling on cases involving species' conservation or thresholds of significance (Offices for project development [16])	–

by politicians and decision-makers) [51]. Yet, scholarly work would grant scientists the ability to communicate and interpret their research findings best [10, 52].

Moreover, the extent of decisive power courts can exercise regarding the interpretation of scientific data is left to question (cf. Project development consultancy [16]). Nevertheless, some actors perceive their role to be arbitrary regarding, for example, the relevance of species for planning (cf. Project development consultancy [16]). This could be interpreted with regard to the "significant increase in mortality", which is a legal term and seems to provide a feeling for insecurity deciding about cases regarding species' conservation (Offices for project development [16]). PROGRESS forced the discussion of how law and court decisions can be better measured (i.e. operationalised) concerning the state of knowledge in ecology, such as a better clarification of loose legal terms (e.g. "significant increase in mortality") into practice (Environmental planning consultancy [36]).

3.2.5 Frame "Climate Protection"

Arguments that dealt with the discrepancy between wildlife and climate protection are outlined in Table 6. Participants in this frame were in particular those actors who have expressed opinions concerning wildlife protection and conservation as well. This frame thus includes these argumentation patterns in the discourse.

Relativists

In the discourse, climate protection is contrasted to wildlife protection and vice versa, while actors struggle for opinion leadership. On the one hand, the issue was perceived on a "higher level", as climate protection was appreciated as an overall measure that attends to every need (Independent network of scientists and parliamentarians [32]; journalists [34], Environmental planning consultancy [36]; Former member of the parliament (Green Party) [40]). Actors argued that in

Table 6 Argumentation patterns within the frame "climate protection" to support or reject PROGRESS' research results

Frame "climate protection"	
+ (research results are valid, mainly referring to population models)	– (research results are not valid or sensitive, mainly referring to fatality searches)
To speed up climate protection at the expense of birdlife is criticised. Birds that should be protected against climate change in the long term would no longer be there (Nature Conservation Association [53])	Renewable energies contribute decisively to climate protection, which in turn contributes decisively to the protection of wildlife (Independent network of scientists and parliamentarians [32]; Environmental planning consultancy [36]; Former member of the parliament (Green Party) [40])

fostering climate protection, species conservation would already be ensured. These actors called for fewer mitigation measures at the local project level. Occurring fatalities or losses were relativised by referring to an overriding public interest.

Change Agents

Otherwise, different actors perceived wildlife protection efforts, e.g. for the red kite and common buzzard, to be an essential component of conservation. This is based on the assumption that the objects to protect would no longer exist to be protected in the long term against the impacts of climate change (Nature Conservation Association [41]). Once again the PROGRESS study was appreciated for signalling that enhanced mitigation measures and adaptations are needed in approval practice in order to fulfil the precautionary principle of environmental law [54].

4 Discussion

The PROGRESS case exemplifies conflicts in science-policy-practice. It indicates a need for approaches to negotiate conflicting perceptions and motives relating to upcoming research findings.

4.1 Participating Actors and Their "Framing" of Wildlife Research

Due to the fact that actors claim leadership of interpretation, conflicting debates are evident. In this case study, it mainly revolves around two key topics: firstly, the different perspectives regarding "wildlife protection and conservation" and, secondly, which "scientific research methods" are able to assess and forecast population effects more accurately for planning and permitting (e.g. telemetry data, Project development and environmental consultancy [30] or modelling like in the PROGRESS case [11–13]). Concerning wildlife protection, the dispute over whose interpretation is correct is particularly rich in its variety – debating mainly the validity of the PROGRESS' results and the interpretation of the results for planning practice. Two arenas of argumentation patterns emerge in fierce opposition to each other: on the one hand, PROGRESS' population models, forecasting future population changes, are deliberately questioned, and, on the other hand, the study's results are accepted, particularly regarding the study's fatality searches.

Thus, an uneasy stalemate has been established in the debate. Actors who hold PROGRESS' research results to be valid generally aim to foster conservation management practices. In contrast, actors who deny aspects of the research findings demonstrate opposite argumentation patterns. It is apparent that matters of wildlife protection and research methods are strongly interlinked. Therefore, the questions arise whether the results derived from particular research methods are "only" triggering the communication of certain motives or whether the discussion surrounding PROGRESS indicates a need for better scientific practice in the wind and wildlife sector [10]. The analysis seems to suggest either way that there are even more actors in the wind energy sector who are denying PROGRESS' research results than accepting them.

Actors typically involved in approval processes for wind energy facilities are represented in the PROGRESS discourse (primarily authorities, conservationists and project developers). While they occupy various standpoints, it is still worth noting that the discussion seems to be politically and economically influenced as well, due to respective actors associated with political parties and with the project development of wind turbines. The processes of relativising, objectifying, changing or preserving the current planning practice represent four contradictory motivations in the PROGRESS discourse, highlighting the question of relevant impacts on sound and objective decision-making processes.

Seemingly, it can be questioned whether any information may become instrumentalised or whether there is actually a need to adjust research approaches. Concerns that scientific research results may be seen through political lenses are not new [55]. Hence, depending on how PROGRESS' research methods are interpreted, be it influenced by external interests, this case may have further ramifications. This applies in particular to the question of how reliable models can be in order to sufficiently forecast population changes.

4.2 *Challenges for Science and Practice*

Against this background, the planning entity as well as the general public is faced with complex problems. Actors seem to mutually weaken the validity of their opponents' arguments by accusing them of representing interest-oriented argumentations and selectively highlighting facts. It is becoming apparent that fixed disputes over whose interpretation of the PROGRESS case is correct cannot guarantee sound decision-making. Mutual fact-checking of opposite argumentation patterns results in a (political) stalemate in the wind energy and conservation sector [36, 41, 56].

It is crucial to understand that the competition for leadership in interpreting research results may exist due to interests to fuel-respective decision-making processes. For instance, proposed arguments are often not evidently and thoroughly grounded in the PROGRESS discourse. More specifically, arguments lack clear references. Long-term stalemate in such discourses may even lead to greater uncertainty in planning and permitting practices, harming both the wind energy and conservation sectors.

4.3 *Actions Needed*

4.3.1 Administrative Leeway and Planning Certainty

The PROGRESS case also makes plain that the statutory planning framework itself leads to discussions. On the one hand, the leeway of the German "assessment prerogative" (German: "Einschätzungsprärogative") in conservation issues seems to be on trial. It is a concept framed by German jurisdiction, allowing authorities to decide, at their discretion, on the merit of each individual case, only if a case is substantiated through lack of secure knowledge in the scientific field of ecology.[3]

On the other hand, the PROGRESS debate sheds a light on the German species protection regulations and practice: Potential effects of wind energy projects must refer to individuals of local populations.[4] Approval authorities must evaluate a threshold of significance in order to determine in which cases local species may be affected by planned projects (is there a "significantly increased collision risk"?). How such thresholds of significance can be determined is currently specific to individual cases and hence constitutes a steadfast element of uncertainty.

However, PROGRESS evaluated possible population effects rather than individual risks. Therefore, it remains to be discussed whether a reference to population effects in legislature is able to supply information on significant effects on species [16]. For example, Bick and Wulfert [57] state that significant effects on species cannot be judged regardless of population biology (e.g. reproduction rates, mortality rates).

4.3.2 Fostered Communication Between Opposing Actors

Even though contrary argumentation patterns and motives are a typical indication, there is a demand for better communication and attention to opposing positions [58]. To prevent further stalemate, it is time to foster exchange between divergent interests [58]. Mutual understanding of perceptions and argumentation patterns is essential, and, thus, scientific debate may also help to clarify issues and to determine the key questions, which need to be addressed in the future [10]. For example, Sullivan et al. [10, p. 11] state that "[...] scientists view the notion that they can openly disagree with one another as a strength of their profession rather than a weakness". Consequently, Sullivan et al. [10, p. 23] underline the role of scientists in assisting the public "[...] with sorting out objective information from highly biased opinion".

In this way, mutual interests as well as knowledge gaps may be addressed in a more effective manner. For example, divergent perceptions in the PROGRESS discourse in the frame "wildlife protection and conservation" make apparent that

[3]cf. BVerwG, judgement of 9 July 2008, 9 A 14.07, paragr. 69.

[4]cf. e.g. BVerwG, judgement of 9 July 2008, 9 A 14/07, paragr. 54.

there is still a lack of detailed knowledge about the efficacy of mitigation measures [59]. Such knowledge is essential, as it even addresses the reliability of the statutory framework. Scientific debate may also address questions on what is finally needed, whether it be on a case-by-case basis or through a metalevel approach [16] to fulfil the aims of species protection.

4.3.3 Enhanced Communication via Best Available Science

Future action is also indicated as far as the communication of new research results is concerned to reduce uncertainties in planning and permitting. As an example, the active discourse framing the prepublication of PROGRESS' results dealt initially with the scope of information and its implications for planning practice [28].

The format of knowledge communication may be key to finding a solution, with a focus on limiting sole leadership in the interpretation and biased selection of data [10, 60]. Questions, such as to what extent scientific information is available but limited as well, and the relevant effects on planning practice, are crucial and must be communicated beforehand [10]. Scientists responsible for such information are perceived to be in the best position to evaluate the scope but also limitations of their data [61]. For example, Mills and Clark [62, p. 189] state that science can facilitate discussion among competing interests to understand management decisions, "[...] when it is properly generated, presented, and accountably used".

In the USA, these subjects are debated under the heading of "best available science" [10, 52, 63, 64]. By means of science mandates, environmental legislation demands the use of "best available science" in the interface between science and policy/practice.[5] This is not only for easier, sound and objective decision-making processes [65, 66]. Scientists are equally advised to communicate correctly the scopes and impacts of their research results on planning practice and executive authority [52, 63]. Fostering best available science offers substantial potential to accentuate the awareness and focus required between science and practice.

5 Conclusion

By identifying actors, and the "framing" of argumentation patterns and typology of motives, it is becoming apparent that discourses in the wind energy and conservation sector are often multi-faceted and highly controversial. In conclusion, the PROGRESS discourse indicates that the publication of new research results – against all expectations – does not necessarily lead to narrowing knowledge gaps and to fostering mutual understanding in planning and permitting procedures. This

[5]cf. "best available science" mandate of Washington State Growth Management Act, RCW 36.70A.172.

is particularly important since it shifts the focus on the dynamic tension inherent in predetermined views on research results – as long as further sound studies make it ever more difficult to argument against repeated evidence.

Apparently, our analysis shows that the ways of communicating research results can influence their perception. This is due to a wide range of interpretation of the very same results and competition for leadership in interpretation and communication. As multiple stakeholders are involved in the wind energy and conservation sector, several motives can be discerned to frame research results. In our case study, we identified four major motives opposing each other (relativists, objectifiers, preservers, change agents).

Further challenges for science and practice emerge. An effort should be made to improve best available science to better respond to the questions at hand. It may help to inform a knowledge society, supposedly less biased by vested interests, thus contributing to more balanced and fair decision-making processes. Necessary pathways encompass also awareness for the question of who might communicate both knowledge and recommendations unanimously within the science-policy-practice interface.

Acknowledgements The analysis was supported by the German Federal Environmental Foundation (Deutsche Bundesstiftung Umwelt). We would like to thank Davy Mains for his help in proofreading the manuscript and in editing the language. We also gratefully acknowledge the comments by two anonymous reviewers on an earlier version of the manuscript.

References

1. MacCoun, R.J.: Biases in the interpretation and use of research results. Annu. Rev. Psychol. (1998). https://doi.org/10.1146/annurev.psych.49.1.259
2. Pullin, A.S., Knight, T.M., Stone, D.A., Charman, K.: Do conservation managers use scientific evidence to support their decision-making? Biol. Conserv. (2004). https://doi.org/10.1016/j.biocon.2003.11.007
3. Cook, C.N., Mascia, M.B., Schwartz, M.W., Possingham, H.P., Fuller, R.A.: Achieving conservation science that bridges the knowledge-action boundary. Conserv. Biol. J. Soc. Conserv. Biol. (2013). https://doi.org/10.1111/cobi.12050
4. Robbennolt, J.K., Studebaker, C.A.: News media reporting on civil litigation and its influence on civil justice decision making. Law Hum. Behav. (2003). https://doi.org/10.1023/A:1021622827154
5. Wayne, W., Guy, G., Cheolhan, L.: Agenda setting and international news: media influence on public perceptions of foreign nations. J. Mass Commun. Q. **81**(2), 364–377 (2004)
6. McGauran, N., Wieseler, B., Kreis, J., Schüler, Y.-B., Kölsch, H., Kaiser, T.: Reporting bias in medical research – a narrative review open access review. Trials. **11**(37), (2010)
7. Rothstein, H.R., Sutton, A.J., Borenstein, M. (eds.): Publication Bias: Recognizing the Problem, Understanding Its Origin and Scope, and Preventing Harm. Wiley, West Sussex (2005)
8. Linhart, E., Dhungel, A.-K.: Das Thema Vermaisung im öffentlichen Diskurs. Z. Agrarpolitk Landwirtsch (2) (2013)
9. Siyal, S.H., Mörtberg, U., Mentis, D., Welsch, M., Babelon, I., Howells, M.: Wind energy assessment considering geographic and environmental restrictions in Sweden. A GIS-based approach. Energy. (2015). https://doi.org/10.1016/j.energy.2015.02.044

10. Sullivan, P.J., Acheson, J.M., Angermeier, P.L., Faast, T., Flemma, J., Jones, C.M., Knudsen, E.E., Minello, T.J., Secor, D.H., Wunderlich, R., Zanatell, B.A.: Defining and implementing best available science for fisheries and environmental science, policy, and management. Report: best science committee. Fisheries. **31**, 460–465 (2006)
11. Grünkorn, T., Rönn, J.v., Blew, J., Nehls, G., Weitekamp, S., Timmermann, H., Reichenbach, M., Coppack, T., Potiek, A., Krüger, O.: Ermittlung der Kollisionsraten von (Greif-)Vögeln und Schaffung planungsbezogener Grundlagen für die Prognose und Bewertung des Kollisionsrisikos durch Windenergieanlagen (PROGRESS), Zusammenfassung. http://bioconsult-sh.de/site/assets/files/1560/1560-1.pdf (2016). Accessed 1 Dec 2017
12. Grünkorn, T., Rönn, J.v., Blew, J., Nehls, G., Weitekamp, S., Timmermann, H., Reichenbach, M., Coppack, T., Potiek, A., Krüger, O.: Ermittlung der Kollisionsraten von (Greif-)Vögeln und Schaffung planungsbezogener Grundlagen für die Prognose und Bewertung des Kollisionsrisikos durch Windenergieanlagen (PROGRESS): Schlussbericht zum durch das Bundesministerium für Wirtschaft und Energie (BMWi) im Rahmen des 6. Energieforschungsprogrammes der Bundesregierung geförderten Verbundvorhaben PROGRESS, FKZ 0325300A-D. http://bioconsult-sh.de/site/assets/files/1561/1561-1.pdf (2016). Accessed 1 Dec 2017
13. Grünkorn, T., Blew, J., Krüger, O., Potiek, A., Reichenbach, M., Rönn, J.v, Timmermann, H., Weitekamp, S., Nehls, G.: A large-scale, multispecies assessment of avian mortality rates at land-based wind turbines in Northern Germany. In: Köppel, J. (ed.) Wind energy and wildlife interactions. Presentations form the CWW2015 Conference, pp. 43–64. Springer (2017)
14. Kreutzfeldt, M.: Windkraft hat einen Schlag bei Vögeln. Eine aktuelle Studie sieht für den Mäusebussard Gefahren durch Windräder. Das könnte Folgen für den Windkraftausbau haben. http://www.taz.de/!5308686/ (2016). Accessed 28 Nov 2017
15. Deutsche WindGuard GmbH: Status des Windenergieausbaus in Deutschland. https://www.wind-energie.de/sites/default/files/attachments/page/statistiken/factsheet-status-windenergieausbau-land-2016.pdf (2016). Accessed 15 Nov 2017
16. Onshore Wind Energy Agency: Windenergie und Artenschutz: Ergebnisse aus dem Forschungsvorhaben PROGRESS und praxisrelevante Konsequenzen. http://www.fachagentur-windenergie.de/fileadmin/files/Veranstaltungen/Diskussions-VA_PROGRESS_17.11.2016/FA_Wind_Programm_PROGRESS_17-11-2016.pdf. (2017). Accessed 1 Dec 2017
17. Wulfert, K., Lau, M., Widdig, T., Müller-Pfannenstiel, K., Mengel, A.: Standardisierungspotenzial im Bereich der arten- und gebietsschutzrechtlichen Prüfung. FuE – Vorhaben im Rahmen des Umweltforschungsplanes des Bundesministeriums für Umwelt, Naturschutz und Reaktorsicherheit im Auftrag des Bundesamtes für Naturschutz – FKZ 3512 82 2100. https://www.bfn.de/fileadmin/BfN/planung/eingriffsregelung/Dokumente/Standardisierungspotenzial_Arten-_und_Gebietsschutz_1.pdf (2015). Accessed 1 Dec 2017
18. Gee, J.P. (ed.): An Introduction to Discourse Analysis: Theory and Method Theory and Method. Routledge, Abingdon (2014)
19. Schiffrin, D., Tannen, D., Hamilton, H.E. (eds.): The Handbook of Discourse Analysis. Blackwell Publishers, Malden (2001)
20. Gee, J.P., Handford, M.: The Routledge Handbook of Discourse Analysis. Routledge (2013)
21. Gamson, W.A., Modigliani, A.: Media discourse and public opinion on nuclear power: a constructionist approach. Am. J. Sociol. **95**(1), (1989)
22. Hess, S., Von Cramon-Taubadel, S., Zschache, U., Theuvsen, L., Kleinschmit, D.: Explaining the puzzling persistence of restrictions on seasonal farm labour in Germany. Eur. Rev. Agric. Econ. **39**, 707–728 (2012)
23. Ruuska, O.: Framing Adaptation to Climate Change – Urban Adaptation to Heat Stress in Three European Capital Cities. Master's thesis. Berlin Institute of Technology (2017)
24. Frantzeskaki, N., Jhagroe, S., Howlett, M.: Greening the state? The framing of sustainability in Dutch infrastructure governance. Environ. Sci. Pol. (2016). https://doi.org/10.1016/j.envsci.2016.01.011
25. Forsyth, D.R.: Moral judgment. Personal. Soc. Psychol. Bull. **7**(2), 218–223 (1981)

26. Bernstein, R.J.: Beyond Objectivism and Relativism: Science, Hermeneutics, and Praxis. University of Pennsylvania Press, Philadelphia (1983)
27. Laudan, L.: Science and Relativism: Some Key Controversies in the Philosophy of Science. The University of Chicago Press, Chicago (1990)
28. Meinhof, R.: Hast du 'nen Vogel? Windkraftgegner lieben den Rotmilan. Denn wo er brütet, darf keine neue Anlage gebaut werden. Energiewende? Über ein plötzlich sehr politisches Tier. http://www.sueddeutsche.de/wissen/tierschutz-hast-du-nen-vogel-1.2805325?reduced=true (2016). Accessed 15 Dec 2017
29. Nisbet, M.C., Scheufele, D.A., Shanahan, J., Moy, P., Brossard, D., Lewenstein, B.V.: Knowledge, reservations, or promise? Commun. Res. (2016). https://doi.org/10.1177/009365002236196
30. Project development and environmental consultancy: Die grössten Fehler der PROGRESS-Studie: für die Ermittlung der Kollisionsraten von (Greif-)Vögeln und Schaffung planungsbezogener Grundlagen für die Prognose und Bewertung des Kollisionsrisikos durch Windenergieanlagen. https://www.hans-josef-fell.de/content/index.php/dokumente/studien-und-analysen/916-die-groessten-fehler-der-progress-studie-stellungnahme-von-dr-oliver-kohle-juli-2016/file (2016). Accessed 1 Dec 2017
31. Project development and environmental consultancy: Windenergie und Rotmilan: Ein Scheinproblem. http://www.pnn.de/fm/61/Windenergie%20und%20Rotmilan_%20Ein%20Scheinproblem%20160125.pdf, https://docs.wixstatic.com/ugd/886e3c_56880b71e62946619b8b8d183f9d7ee3.pdf (2016). Accessed 1 Dec 2017
32. Independent network of scientists and parliamentarians: Windkraft gegen Vögel? Progress Studie gibt Aufschluss. http://w3.windmesse.de/windenergie/news/22436-windkraft-gegen-vogel-progress-studie-gibt-aufschluss (2016). Accessed 14 Dec 2017
33. Haerdle, B.: Wir sehen die Ergebnisse kritisch: Interview. http://www.umweltbriefe.de/docs/ub_15_16_dezember/naturschutz.html (2016). Accessed 16 Dec 2017
34. Weinhold, N.: Neuer Problemvogel für die Windkraft. https://www.erneuerbareenergien.de/neuer-problemvogel-fuer-die-windkraft/150/434/92551/ (2016). Accessed 15 Dec 2016
35. Schreiber, M., Langgemach, T., Dürr, T.: Hoher Aufwand, vage Resultate: Windenergie und Vogelschutz – Anmerkungen zur Progress-Studie. Naturschutz Landschaftsplanung (NuL). **10**, 328–332 (2016)
36. Environmental planning consultancy: Aktuelle Fragen des Natur- und Artenschutzes. BWE Seminar "Natur- und Artenschutz – Neuigkeiten für die Windparkplanung", Berlin (2017)
37. Haerdle, B.: Fortschritt ins Ungewisse: Greifvögel vs. Windkraft. http://www.umweltbriefe.de/docs/ub_15_16_dezember/naturschutz.html (2016). Accessed 16 Dec 2017
38. A ministry: Windkraftanlagen und Greifvögel: Fragen und Antworten zu aktuellen Forschungsvorhaben. http://www.bmub.bund.de/themen/natur-biologische-vielfalt-arten/artenschutz/vogelschutz/windkraftanlagen-und-greifvoegel/ (2017). Accessed 15 Dec 2017
39. Society for the Conservation of Owls: Kommentar zur PROGRESS-Studie. http://www.ipola.de/sites/default/files/NEUES_anhaenge/Breuer_EGE_2016_Kommentar%20zur%20PROGRESS-Studie.pdf (2016). Accessed 1 Dec 2017
40. IWR: Was der Windenergieausbau für Rotmilan und Mäusebussard bedeutet. http://www.iwr.de/news.php?id=31731 (2016)
41. Nature Conservation Association: Windenergie-Lobby leugnet Artenschutzproblematik. https://www.nabu.de/news/2016/06/20834.html (2016). Accessed 14 Dec 2017
42. Richarz, K.: Windenergie im Lebensraum Wald: Gefahr für die Artenvielfalt. Situation und Handlungsbedarf. https://www.deutschewildtierstiftung.de/publikationen (2016)
43. Project development consultancy: PROGRESS-Studie gibt keinen Aufschluss über Auswirkung der Windkraft auf geschützte Vogelarten. https://www.abo-wind.com/de/aktuelles/pressemitteilungen/2017/2017_01_02_Progress_Maeusebussard.html (2017). Accessed 16 Dec 2017
44. Hötker, H., Krone, O., Nehls, G.: Greifvögel und Windkraftanlagen: Problemanalyse und Lösungsvorschläge: Schlussbericht für das Bundesministerium für Umwelt, Naturschutz und Reaktorsicherheit (BMU). https://www.nabu.de/downloads/Endbericht-Greifvogelprojekt.pdf (2013). Accessed 1 Dec 2017

45. Reichenbach, M., Brinkmann, R., Kohnen, A., Köppel, J., Menke, K., Ohlenburg, H., Reers, H., Steinborn, H., Warnke, M.: Bau- und Betriebsmonitoring von Windenergieanlagen im Wald: Abschlussbericht 30.11.2015, Erstellt im Auftrag des Bundesministeriums für Wirtschaft und Energie. http://www.arsu.de/sites/default/files/projekte/wiwa_abschlussbericht_2015.pdf (2015). Accessed 1 Dec 2017
46. Biologische Station Kreis Paderborn: Ergebnisbericht zur Erfassung des Rotmilanbestands im Kreis Paderborn. http://www.bs-paderborn-senne.de/fileadmin/user_upload/downloads/Ergebnisbericht_Rotmilan_2016_Text.pdf (2016). Accessed 1 Dec 2017
47. Weinhold, N.: Sechs neue Erkenntnisse zu Vogelschutz und Windkraft. https://www.erneuerbareenergien.de/bwe-artenschutz-debatte-versachlichen/150/434/96611/ (2016). Accessed 15 Dec 2016
48. Wind energy association: PROGRESS-Endbericht wird Debatte um Artenschutz versachlichen. https://www.wind-energie.de/presse/pressemitteilungen/2016/progress-endbericht-wird-debatte-um-artenschutz-versachlichen (2016). Accessed 20 Nov 2017
49. Working Group of German State Bird Conservancies (LAG VSW): Recommendations for distances of wind turbines to important areas for birds as well as breeding sites of selected bird species (as at April 2015). http://www.vogelschutzwarten.de/downloads/lagvsw2015.pdf (2015). Accessed 1 Dec 2017
50. Gellner, T., Pauly, B.: Volksbegehren startet – Windräder bedrohen Brandenburgs Wappentier. http://www.maz-online.de/Brandenburg/Windraeder-bedrohen-Brandenburgs-Wappentier (2016). Accessed 15 Dec 2017
51. Krumenacker, T.: Windenergie und Mäusebussard: Wir haben eine potenziell bestandsgefährdende Entwicklung. https://www.nabu.de/imperia/md/content/nabude/vogelschutz/der_falke_progress-artikel_m__rz16.pdf (2016). Accessed 20 Nov 2017
52. Wolters, E.A., Steel, B.S., Lach, D., Kloepfer, D.: What is the best available science? A comparison of marine scientists, managers, and interest groups in the United States. Ocean Coast. Manag. (2016). https://doi.org/10.1016/j.ocecoaman.2016.01.011
53. Crolly, H., Wetzel, D.: "Signifikant erhöhtes Tötungsrisiko": Ist die Energiewende wichtiger als der Schutz bedrohter Arten wie des Milans? Ein Windkraft-Projekt im Süden könnte zum Präzedenzfall werden. https://www.welt.de/print/wams/wirtschaft/article157779800/Signifikant-erhoehtes-Toetungsrisiko.html (2016). Accessed 1 Dec 2017
54. UNCED: The Rio Declaration, Principle 15 – the Precautionary Approach. (1992). http://www.gdrc.org/u-gov/precaution-7.html. Accessed 16 Dec 2017
55. Rainie, L., Funk, C., Kennedy, B., Anderson, M., Duggan, M., Olmstead, K., Perrin, A., Greenwood, Suh, M, Porteus, M., Page, D.: Americans, Politics and Science Issues. http://assets.pewresearch.org/wp-content/uploads/sites/14/2015/07/2015-07-01_science-and-politics_FINAL-1.pdf (2015)
56. Former member of the parliament: Pressemitteilung zum NABU – Faktencheck "Rotmilan und Windenergie" wissenschaftlich nicht haltbar. https://www.hans-josef-fell.de/content/index.php/presse-mainmenu-49/infobriefe-mainmenu-72/975-infobrief-12-2016-stellungnahme-zum-nabu-faktencheck-rotmilan-und-windenergie-podiumsdiskussion-fluchtursachen-buergerbegehren-raus-auf-der-steinkohle (2016). Accessed 14 Nov 2017
57. Bick, U., Wulfert, K.: 40. Umweltrechtliche Fachtagung vom 10. bis 12. November 2016: Artenschutzrecht in der Vorhabenzulassung. http://www.gesellschaft-fuer-umweltrecht.de/wp-content/uploads/2017/02/Tagung-40_2016_Tagungsmappe.pdf (2016). Accessed 1 Dec 2017
58. Kuhn, D., Udell, W.: Coordinating own and other perspectives in argument. Think. Reason. (2007). https://doi.org/10.1080/13546780600625447
59. Gartman, V., Bulling, L., Dahmen, M., Geißler, G., Köppel, J.: Mitigation measures for wildlife in wind energy development, consolidating the state of knowledge — part 1. Planning and siting, construction. J. Environ. Assess. Policy Manag. (2016). https://doi.org/10.1142/S1464333216500137
60. Peters, H.P.: Gap between science and media revisited. Scientists as public communicators. Proc. Natl. Acad. Sci. U.S.A. (2013). https://doi.org/10.1073/pnas.1212745110

61. Brossard, D., Nisbet, M.C.: Deference to scientific authority among a low information public. Understanding U.S. opinion on agricultural biotechnology. Int J Publ Opin. Res. (2006). https://doi.org/10.1093/ijpor/edl003
62. Mills, T.J., Clark, R.N.: Roles of research scientists in natural resource decision-making. For. Ecol. Manag. (2001). https://doi.org/10.1016/S0378-1127(01)00461-3
63. Ryder, D.S., Tomlinson, M., Gawne, B., Likens, G.E.: Defining and using 'best available science'. A policy conundrum for the management of aquatic ecosystems. Mar. Freshw. Res. (2010). https://doi.org/10.1071/MF10113
64. Charnley, S., Carothers, C., Satterfield, T., Levine, A., Poe, M.R., Norman, K., Donatuto, J., Breslow, S.J., Mascia, M.B., Levin, P.S., Basurto, X., Hicks, C.C., García-Quijano, C., St. Martin, K.: Evaluating the best available social science for natural resource management decision-making. Environ. Sci. Pol. (2017). https://doi.org/10.1016/j.envsci.2017.04.002
65. Murphy, D.D., Weiland, P.S.: The route to best science in implementation of the Endangered Species Act's consultation mandate: the benefits of structured effects analysis. Environ. Manag. (2011). https://doi.org/10.1007/s00267-010-9597-9
66. U.S. Department of Commerce, National Oceanic and Atmospheric Administration (NOAA), National Marine Fisheries Service: Technical Guidance for Assessing the Effects of Anthropogenic Sound on Marine Mammal Hearing. Underwater Acoustic Thresholds for Onset of Permanent and Temporary Threshold Shifts. NOAA Technical Memorandum NMFS-OPR-55. http://www.nmfs.noaa.gov/pr/acoustics/Acoustic%20Guidance%20Files/opr-55_acoustic_guidance_tech_memo.pdf (2016). Accessed 1 Dec 2017

Estimating the Potential Mortality of Griffon Vultures (*Gyps fulvus*) Due to Wind Energy Development on the Island of Crete (Greece)

Stavros M. Xirouchakis, Efi Armeni, Stamatina Nikolopoulou, and John Halley

Abstract Crete has been characterized as an area with a high wind energy capacity due to its mountainous terrain and the strong prevailing winds throughout the year. At the same time, the island constitutes the last stronghold for vulture species in Greece, currently holding the largest insular population of Eurasian griffons (*Gyps fulvus*) worldwide (ca. 1000 individuals). Given the empirical data on the mortality of large raptors due to collisions with wind turbine blades, the aim of the present study was to predict the potential impact of wind energy installations on the griffon vulture population on the island. The study was developed in two steps, namely, (a) the spatial mapping of the existing and planned wind energy projects up to the year 2012 and the delineation of their risk area and (b) the calculation of the annual collision rate based on the expected number of vulture risk flights and the probability of being killed. Overall, the minimum number of fatalities due to collision of vultures to wind turbines was estimated at 84 individuals per year. However, this figure could drop by over 50% if the European network of the NATURA 2000 sites was set as an exclusion zone for wind energy facilities. The study pinpoints the need for proper siting of wind farms and the prerequisite of sensitivity mapping for vulnerable species prone to collision on wind turbines.

S. M. Xirouchakis
Natural History Museum of Crete, University of Crete, Heraklion, Greece

School of Science & Engineering, University of Crete, Heraklion, Greece
e-mail: sxirouch@nhmc.uoc.gr

E. Armeni
Biology Department, University of Crete, Heraklion, Greece

S. Nikolopoulou
Natural History Museum of Crete, University of Crete, Heraklion, Greece
e-mail: snikolo@hcmr.gr

J. Halley
Department of Biological Applications & Technology, University of Ioannina, Ioannina, Epirus, Greece
e-mail: jhalley@cc.uoi.gr

R. Bispo et al. (eds.), *Wind Energy and Wildlife Impacts*,
https://doi.org/10.1007/978-3-030-05520-2_13

Keywords Wind farm · Collision rate · Cumulative impact · NATURA 2000 site network

1 Introduction

The environmental and socio-economic repercussions of extracting, trading and consuming fossil fuels and gas emissions have led societies to re-evaluate current energy policies and consider lower-impact forms of power generation. Renewable energy sources have received substantial government support as they do not contribute to global warming [1]. The European Commission has set a target of 20% of the EU energy to be generated though renewable resources by 2020 (EU Directive 2001/77/EC) with wind farms being one of the main alternative energy sources mitigating greenhouse emissions [2]. However, the environmental impact of commercial wind power production on biodiversity has proved to be substantial [3–7]. Wildlife is affected by wind power production through habitat loss, disturbance and displacement and above all by increased collision risk with wind turbines [8–10]. Bird fatalities due to collision with wind turbines have been the most prominent and frequently identified environmental drawback of wind energy development. Bird casualties from collisions can reach up to 40 deaths per turbine per year [11] with large raptors suffering the greatest toll [12–17]. However, there are also studies that have produced very low figures of bird strikes in wind farms from 0.001 to 0.06 birds/turbine/year [18, 19] indicating that mortality rates are highly variable depending on the technical specification and arrangement of wind turbines, local weather conditions, landscape topography and the flight behaviour of the species involved [18, 20–24].

The European Commission has produced a rather well-designed legislative and consulting framework for the anticipated development of wind energy facilities taking into consideration aspects of nature conservation. This includes the application of Environmental Strategic Assessment (ESA) process and proper Environmental Impact Assessment (EIA) studies targeted to minimize negative impacts on wildlife [25, 26]. The implementation of proper spatial planning and the development of guidance documents [27] also try to make renewables compatible with nature conservation. However, the experience acquired shows that the adverse effects of wind energy facilities on wildlife are site- and species-specific. Sensitivity mapping, i.e. developing maps that project the possible impact of wind energy installations on wildlife as a function of their geographical location, could be a useful tool for site planning at the stage of strategic design. These maps could relax the pressure of wind energy projects on species or sites of EU conservation concern, e.g. Annex I bird species or a NATURA 2000 site under the provisions of the "Birds" (147/2009/EC) and the "Habitats" (92/43/EC) Directives [28, 29].

In Greece, over 50% of renewable energy is due to come from wind power. To date, locating suitable areas to install wind farms has been based primarily on generation potential, with a priority on islands and mainland mountainous regions resulting in the proliferation of wind energy plans across the country. Even though an Environmental Impact Assessment is required prior to a wind farm installation (EIA Directive), careful analysis of documentation and environmental monitoring has shown that studies on the cumulative impact of wind farms on bird populations are absent and regional sensitivity mapping has not been elaborated at all [30, 31]. The island of Crete with its mountainous terrain and its geographical position at south Aegean holds a high wind energy potential. At present there is an installed power of over 190 MW though there are still more than 3000 MW under plan or in licensing process. However, at the same time, the island is the last stronghold of several threatened European avian species [32, 33]. In the present study, we investigate the cumulative impact of the planned wind energy facilities on Crete focusing on a large raptor with a wide distribution and high local abundance. We evaluated the consequences of wind farm development on the griffon vulture (*Gyps fulvus*) which was regarded as a suitable model species. Griffons are among the most collision-prone large soaring raptors and perhaps the most frequent victims of turbine blades in the Mediterranean region, i.e. up to 1.88 individuals/ turbine/ year [8]. Furthermore, assuming that the most crucial factor in minimizing the negative impact of wind farms on wildlife should be proper siting, we tried to estimate the potential collision mortality of the species by taking into account the existing and all planned wind energy projects on Crete. We applied a collision risk model (CRM) on two different scenarios: (1) the wind power potential is the only criterion for the site planning of wind energy facilities, and the entire island is available for their development, and (2) the NATURA 2000 site network of protected areas is set as an exclusion zone for wind farming. This work constitutes a major step towards the need for a proper sensitivity mapping process for sites of high conservation interest in Crete and could facilitate a comprehensive spatial management approach to the rapidly expanding wind farm industry in Greece.

2 Material and Methods

2.1 Study Area and Species

The study area covered the whole of Crete (35°45′–34°45′N, 23°30′–26°30′E; 8261 km^2) over 256 km from east to west and on average about 40 km from north to south (range 11–56 km). The island is characterized by a rugged terrain (60%) and large mountains of up to 2450 m a.s.l. with a few plains and many of hills covered with extensive vineyards and olive groves. The climate possesses a strong regional gradient with annual rainfalls of over 2000 mm in the northwest and less than 300 mm in the southeast. The mean annual temperature varies between 8 and 20°C [34]. Natural vegetation has been modified by humans for at least the last 3000 years and at present is dominated by cushion-type shrubs called phrygana, dry-leafed maquis and sparse forests and patches of Calabrian pine (*Pinus brutia*), kermes oak (*Quercus coccifera*), cypress (*Cupressus sempervirens*) and oaks (*Quercus pubescens, Quercus macrolepis*). Griffon vultures once widely distributed in continental Greece and some islands [35] have decreased dramatically due to the decline of transhumance pastoralism, illegal shooting and particularly the use of poisoned baits in antipredator campaigns during the 1960s [33, 36]. Presently the species has been listed in the Greek Red Data Book as "endangered" in the mainland and "vulnerable" for the island of Crete [37]. Crete holds the last healthy population in the country (ca. 1000 individuals) which constitutes the largest indigenous insular population worldwide (Xirouchakis unpublished data). The species total area of occurrence has been estimated at 7000 km^2 (84% of Crete) with seasonal differences in correspondence to transhumant livestock [38]. The study period covered the breeding season of 2012–2013 where all breeding colonies and communal roosts were surveyed in the framework of annual monitoring.

2.2 *Data Collection and Statistical Analysis*

Considering that the griffon vulture is a central-place forager, we adopted an "agent"-based modelling process [39] taking into account (1) the location of griffon breeding and permanent roosting cliffs, (2) their population size (number of individuals concentrated in roosts and colonies), (3) the distribution of wind farm facilities that are operational or under plan (wind farm polygons and the turbines therein) and (4) the landscape (in pixels) that constitutes the species foraging habitat. The annual mortality of griffons due to lethal interactions with wind turbines was calculated by an impact function that quantified the spatial overlap of wind farms with the home range of individual colonies and roosts and combined with the seasonal foraging movements of the species and a collision risk model (CRM). For the latter we implemented the "Band CRM" produced by Band et al. [40] which calculates the probability of a bird passing through the rotor swept volume being fatally hit by the blades of the wind turbines. The model incorporates the spatial structure and operation of the wind farm, the technical specification of the turbines and the biometrics and flight characteristics of the target species. It is mathematically robust and has been widely used in Environmental Impact Assessment studies [41–43]. Nevertheless, some key parameters must be predetermined or estimated accurately by survey data, namely, the rotor diameter, the rotation speed, the flight speed and the avoidance rates of the birds. The parameters used in the present case (see Appendix) were incorporated in the procedure of the collision risk model by the development of the following steps:

1. The model's time frame was set to be one day so the number of dead griffon per day would be calculated by multiplying the number of vultures passing through the rotors daily (stage 1) by the probability of an individual being hit (stage 2). In the meantime, the study area was divided into 5×5 km landscape pixels (i.e. grid cells) superimposed on the entire island and consequently on polygons of operational and proposed wind facilities. Spatial data and technical characteristics of wind farms (e.g. polygon coordinates, polygon area, number and type of wind turbines, etc.) were obtained from the web portal

of the Regulatory Authority for Energy in Greece [44] for the study period (Fig. 1), whereas the technical specifications regarding the wind turbines were those retrieved in the RAE geodatabase (http://www.rae.gr/site/portal.csp) and described in the manufacturers' websites.

2. Griffon movements were differentiated by two seasonal periods as defined in respect to the duration of rainfalls and the snow cover on mountainous areas (>800 m), i.e. winter lasting 150 days (15 November–14 April) and summer lasting 210 days (15 April–14 November) [38]. The species daily foraging budget in Crete varies from 8.4 h in summer to 6.8 h in winter, while its flight speed (f_v) has been calculated at 43 km/h. Its home range (h_p) has been determined at 692 km^2 [45]. The latter figure is equivalent to 28 landscape pixels meaning that the foraging sortie of a griffon covers 58.5 and 72.2 pixels ($n_p = f_d/5$ km) around its colony (or roost) during winter and summer, respectively, or an individual vulture might prospect daily each pixel within its home range twice in winter and three times in summer ($R_{pc} = n_p/h_p$).
3. Griffon vulture colonies were spatially mapped and georeferenced in the Hellenic geodetic reference system. Based on distribution patterns of the species (Fig. 2), the relative presence of griffons per pixel was calculated as $p_i = n_i/N_{hj}$, where n_i is the number of griffon observations in each pixel and N_{hj} is the colony-/roost-specific population size, i.e. average number of vulture counts.
4. The number of griffons' risky flights within wind farms (stage 1) was calculated by the frequency individual birds visited each pixel within their colony's home

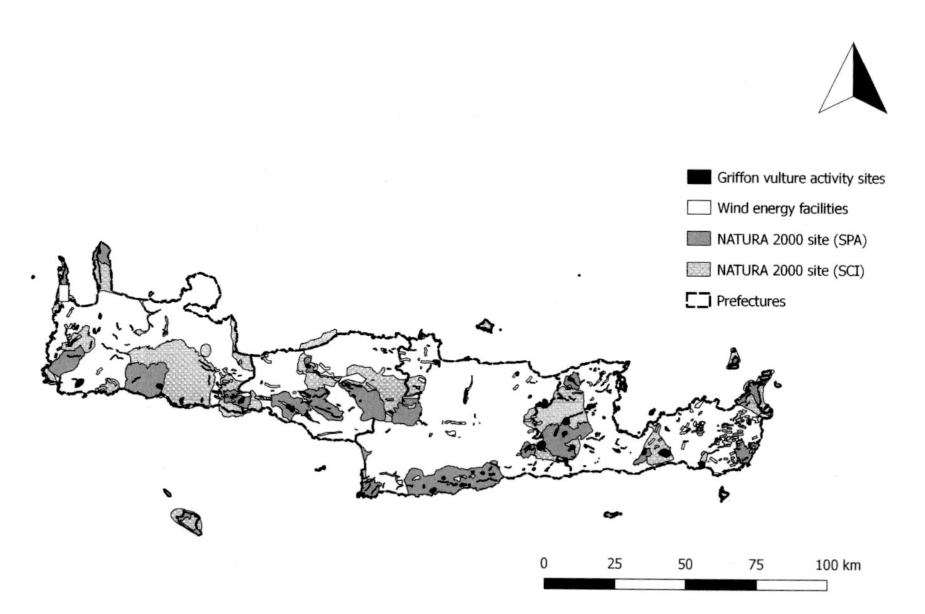

Fig. 1 Wind farm polygons, griffon vulture activity sites (i.e. breeding colonies and permanent roosts) and the NATURA 2000 site network, i.e. Special Protection Areas (SPA) and Sites of Conservation Interest (SCIs) in Crete (prefectures from east to west: Chania, Rethymnon, Heraklion, Lasithi)

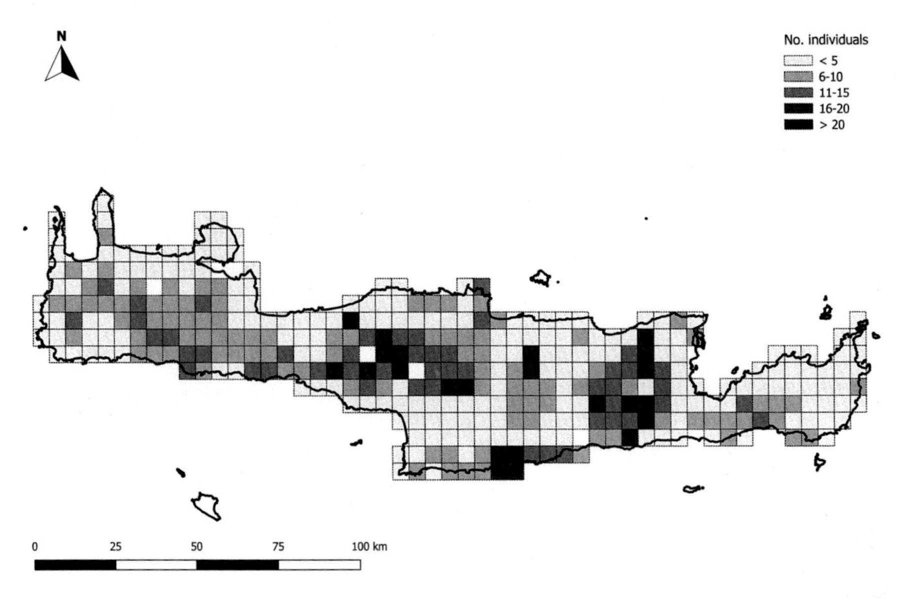

Fig. 2 Local density and total area of occurrence of griffon vultures in Crete mapped in 25 km^2 grids based on the mean number of individuals observed per grid ($\bar{x}$ summer + $\bar{x}$ winter/2) [38]

range (R_{pc}). More precisely, the total distance a vulture may travel during a day was given by the equation $f_d = f_t * f_v$ where f_t was daily foraging budget and f_v was its flight speed. Foraging duration (f_t) differs depending on the photoperiod, which determines the available foraging time (i.e. visibility and the time needed for the formation of suitable thermals and air currents used by foraging griffons).

5. For a collision to take place, a vulture should both be within a wind farm polygon and fly at the height of the rotor blades striking on them if it flies through the rotor swept area without avoiding it. These conditions were considered by working out a correction factor for the number of visits to a pixel (C_f for value R_{pc}). To determine the number of risky flights, we calculated the ratio of wind farm polygons to pixel area. We implemented each stage of the CRM for the wind farm polygons (or portions of them) within in each pixel as the relative presence of griffons in each pixel was different in relation to the seasonal distribution pattern of the species. We then used a simplified general vulture flight model, including the technical characteristics of the turbines at each wind farm. We considered that the geometry of vulture flight resembles a right-angled triangle while the flying height was fixed at 190 m, i.e. the median value in relation to ground altitude (unpublished data from two griffons radio-tagged with GPS/GSM transmitters with barometer sensors). The resulting correction factor was $C_f = (h/H) * (aw/AT)$ where h = turbine height, H = median vulture flight height, a_w = wind farm polygon area and A_T = pixel area (25 km^2).
6. Collision probability (P_{col}), namely, the probability of an individual being hit when passing through a rotor (stage 2), was calculated assuming that birds are

cruciform (cross-shaped) and that turbine blades have no width. Only turbine rotor collisions were considered, assuming that every collision results in fatality. In addition, wind farm area was calculated assuming a straight-line array of turbines [41]. An avoidance rate (A_v) was also included in the mortality function as most raptors actively bypass wind farms [40, 46]. Relevant figures had been estimated empirically by monitoring operating wind farms in the study area [47] and constitute the worst-case scenario with a probability level of 75%. The morphometric values used in the model were those provided for griffon vultures on Crete, i.e. body weight 7.6 kg, wing span 2.56 m and wing surface 0.880 m^2 [48].

7. The potential number of collisions per day for an individual within the home range of its colony was derived by the function:

$$\sum_{i=1}^{28} Q = (\mathrm{R_{pc}} * \mathrm{C_f}) * \mathrm{P_i} * \mathrm{P_{col}} * (1 - \mathrm{A_v})$$

where:

Q is the daily mortality of an individual within the home range of its colony.
i is the number of pixel in the home range of a colony where a wind farm is operating or under plan.
R_{pc} is the number of times a vulture visits each pixel in the home range of its colony.
C_f is the correction factor for the number of risky flights of griffons within a wind farm.
P_i is the relative presence of individuals from each colony per pixel.
P_{col} is the collision probability.
A_v is the avoidance probability.

The potential number of collisions of all the individuals occupying a colony or roost per day was calculated by using the equation $D_j = N_j * Q$, where N_j is the number of individuals (i.e. group size) and Q the mortality of an individual within the home range of its colony/roost, over the course of a single day. The number of deaths in winter and summer was calculated by taking the results of the previous equation and multiplying by the number of days in each season, namely, $\mathrm{Dw}_j = D_j * N_w$, where D_j is the daily number of collisions by all the individuals of a colony/roost and N_w the winter period (in days), and similarly for summer $\mathrm{Ds}_j = D_j * N_s$ where D_j is the daily number of collisions by all the individuals in a colony/roost and N_w summer period (in days). The annual cumulative mortality for each colony/roost was then estimated as $D_{tot} = \mathrm{Dw}_j + \mathrm{Ds}_j$.

Data normality was checked by the use of the Shapiro-Wilk test. Then, under the two wind farming scenarios (i.e. #1, no exclusion zones; #2, NATURA 2000 site network excepted), the mean number of griffon fatalities between the island's prefectures as well as between different colony size classes was investigated by applying non-parametric Kruskal-Wallis post hoc multiple comparison tests [49]. Similarly a Spearman rank correlation and a linear regression were carried out in order to test the relationship between griffon colony size and the predicted number of griffon casualties due to collision to wind turbines [49].

3 Results

A total of 56 active griffon colonies were located and monitored on the island during the study period. They hosted ca. 760 individuals of which 468 consisted of mature birds (234 breeding pairs). Additionally, 17 winter and 16 summer communal roosts were detected which were regularly used by a mean total of 174 and 121 individuals, respectively. Average colony size was calculated at 15 ± 12 individuals (range = 2–50). The wind energy facilities on the island (operational or under plan) accounted for 487 wind farm polygons that incorporated ca. 2900 wind turbines. Overall 32% of these wind farms and 68.6% of the anticipated wind power were placed within the NATURA 2000 site network occupying 9% of its total area. In the meantime, the network hosted 50% of the griffon vulture colonies and 40.7% of the species population (i.e. number of individuals) on the island.

The model predicted that 39% of the griffon colonies which were occupied by more than 15 individuals would account for 62% of the wind farms and vulture interactions and would suffer 65% of the expected mortalities. The overall collision mortality rate was estimated at 0.03 vultures/wind turbine/year producing an annual loss ranging from 3.7% to 11% of the species population. More specifically a total of 990 individuals were estimated to be at threat of striking with turbine blades. The scenario #1 predicted a mean annual mortality of 1.49 ± 1.12 individuals (range = 0.18–4.98) per colony, whereas the overall annual fatality was anticipated at 83.5 griffons. Likewise, under scenario #2 the mean annual mortality per colony was estimated at 0.50 ± 0.47 individuals (range = 0.01–2), but the estimated annual fatality dropped by over 50% to 28.5% griffons. Most of the griffon casualties (ca. 40%) were anticipated in the central part of the island (i.e. prefecture of Heraklion) where high concentrations of vultures and overlapping home ranges of neighbouring colonies occur (Table 1). In both scenarios the differences in the mean annual fatality between the four prefectures of the island were statistically

Table 1 Number of griffons interacting with wind energy facilities and estimated total number of fatal collisions per year (CRM with 75% avoidance rate) under scenario #1 (i.e. no exclusion zones for wind energy facilities) and scenario #2 (NATURA 2000 site network excluded for wind farming) (in parenthesis mean ± s.d. and range of casualties per griffon colony)

Prefecture	No. of individuals	No. of fatalities (scenario #1)	No. of fatalities (scenario #2)
Chania	189	11.4 (0.87 ± 0.49, range = 0.3–1.9)	2.9 (0.22 ± 0.25, range = 0.3–1.9)
Rethymnon	203	25.7 (2.34 ± 1.19, range = 1.4–4.5)	7.2 (0.65 ± 0.45, range = 0.3–1.9)
Heraklion	438	32.7 (1.55 ± 1.19, range = 0.2–4.4)	11 (0.52 ± 0.47, range = 0.3–1.9)
Lasithi	160	13.7 (1.24 ± 1, range = 0.2–2.5)	7.3 (0.66 ± 0.6, range = 0.3–1.9)

Table 2 Population size (no. individuals) of griffon vulture colonies and estimated total number of fatal collisions in wind energy facilities positioned within their home range (in parenthesis mean ± s.d. and range of fatal collisions per colony)

Colony size (*n*)	No. of fatalities (scenario #1)	No. of fatalities (scenario #2)
0–15 (34)	29.1 (0.85 ± 0.5, range = 0.2–2)	10.7 (0.31 ± 0.28, range = 0.01–1)
16–30 (16)	33 (2 ± 0.84, range = 1–3.8)	14.3 (0.89 ± 0.57, range = 0.10–2)
>30 (6)	21.4 (3.5 ± 1.1, range = 1.8–4.9)	3.5 (0.57 ± 0.45, range = 0.09–1.3)

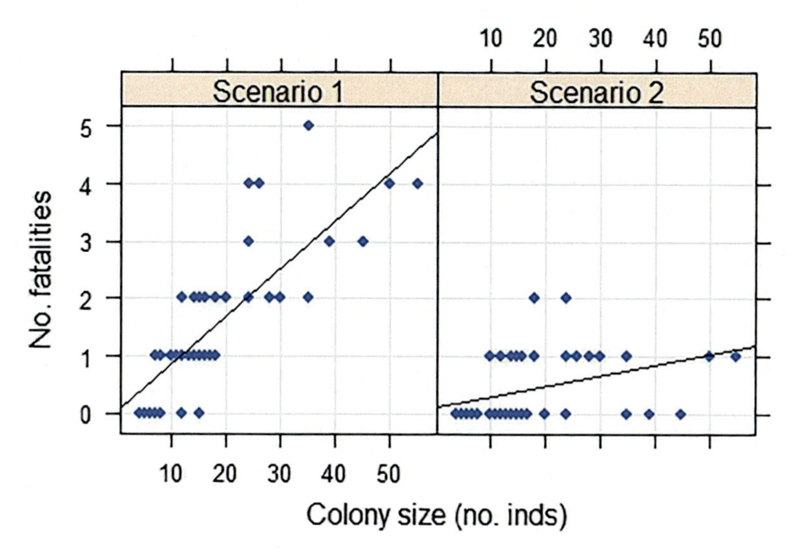

Fig. 3 Relationship between griffon colony size and the expected number of wind farm fatalities expressed by regression functions' two scenarios: (1) no exclusion or buffering zone around griffon activity sites, i.e. breeding colonies and communal roosts, (2) NATURA 2000 site network excluded from installation of wind energy facilities

significant (Kruskal-Wallis test, #1: $H_3 = 11.7$, $P = 0.008$; #2: $H_3 = 8.2$, $P = 0.04$). Under the scenario #1, the large and medium colonies were equally affected by wind farms. The mean number of griffons dying due to collision to wind turbines per year differed significantly from the one predicted for the smaller ones (Kruskal-Wallis post hoc multiple comparison test, $H_2 = 31.8$, $P = 0.001$, Table 2) (Fig. 3). Under the scenario #2, the mean annual mortality of griffons was different only between small and medium colonies with the latter being significantly more affected (Kruskal-Wallis post hoc multiple comparison test, H2 = 15.1, $P = 0.001$, Table 2). Overall there was a positive association between colony size and the number of vultures predicted to be fatally hit by wind turbines which was more pronounced when no wind farming exclusion zone was set (scenario #1: linear regression, $F_{1,54} = 117.6$, $P < 0.001$ and $r_s = 0.82$ vs. scenario #2: $F_{1,54} = 13.6$, $P < 0.001$ and $r_s = 0.48$, $P < 0.001$, Fig. 3).

4 Discussion

Wildlife-renewables' interactions have highlighted three broad adverse effects, namely, disturbance, displacement and direct mortality, due to collision with wind turbines [50]. In the latter case where the impact is most prominent, bird fatalities due to wind turbines are found to vary considerably from one wind farm to another, due to a multitude of factors [8, 18]. So far, most studies have focused on the distance of sensitive areas from wind farms and the concentration points of wildlife based on the notion that collision mortality is elevated near breeding or roosting sites and movement corridors [15, 22, 43, 51]. In Greece competent authorities evaluate energy projects before reaching to a decision under the national and EU legislation (e.g. 97/11/EC Directive) primarily focusing on priority habitats (under the 92/43/EC "Habitats" Directive) and Annex I bird species (under the 147/2009/EC "Birds" Directive). Most Environmental Impact Assessment studies are based on searches for critical areas for birds on the wind farm construction sites and on identifying and mapping large soaring species observed from specific vantage points during particular time of the year (e.g. breeding season, migration periods, etc.). However, in the case of griffons, it has been suggested that local abundance around wind turbines might increase collision risk rates and would be more influential than just the distance of their breeding or roosting cliffs from wind farms [52]. On the other hand, the vultures' spatial distribution within risk areas (i.e. wind farm polygons) could be used as a sound criterion for large-scale environmental planning. Furthermore, modelling the flight behaviour of soaring raptors possesses a high predictive power and has been proposed as a significant management tool for strategic site planning [10, 53, 54].

In the present study, we attempted for the first time to assess the potential cumulative impact of the wind farm industry on the griffon vulture on Crete. We took into consideration the species entire population and predicted the fatalities caused by both the existing and proposed wind farm facilities on the island. We elaborated a spatial model that simulated the foraging distance of the species around its breeding colonies and communal roosts and determined the cumulative collision risk by building an impact function that involved the species space use, group sizes and avoidance rate which affects the outcome in almost any case study. In fact, avoidance rate is usually the most influential parameter in CRMs including the "Band" model [16, 46, 55]. Small variations in its values can dramatically change the prediction of bird fatalities. In the existing literature, avoidance rates are usually set very high (e.g. 99% [46]) based on empirical data associated with weather, light conditions and the topography of area as well as the flight skills of the species at risk [46, 55]. In our study, avoidance rates were estimated in situ in a wind farm where environmental monitoring has been carried out for a full year [47] leading to an avoidance rate of 75%. At the same installation, we accounted for vultures' risky flights, i.e. at rotor height ($n = 1154$), and identified those birds that actively changed flight direction circumventing the wind farm polygon. We believe that this figure was a reasonable approximation of the worst situation which is especially so during the post-fledging dispersal period of young griffons (autumn) and during foggy weather when most fatalities occur [47, 56, 57]. Avoidance rates of griffons might easily close to 75–80% because of their low manoeuvrability and certain limitations associated with this species eyesight and how it perceives visual information. Griffons frequently employ lateral vision for locating conspecifics and foraging opportunities which is more important than looking ahead in open airspace [58]. Besides the rather low avoidance rate used in the present study, was counterbalanced by the low figure used for griffon crossings (i.e. 2–3) per pixel with wind farm per day which is substantially smaller than the actual ones in operating wind farms [47, 56, 57].

Estimating the cumulative impact of wind energy facilities would assist policy and decision makers to combat direct mortality of vultures from wind energy development at a regional scale and minimize their impact on the entire raptor community on Crete which consists mainly of central-place foragers (i.e. bearded vulture (*Gypaetus barbatus*), golden eagle (*Aquila chrysaetos*) and the Bonelli's eagle (*Aquila fasciata*)). Proper regional site planning at the early stages of designing wind energy plants would minimize compensatory measures in the future or the need to relocate or power down problematic wind turbines during critical periods of the day or the year [43, 59]. In this way, the cumulative impact of certain wind energy projects will be properly and adequately assessed. This is especially true when large wind farms are deliberately divided into many small ones in an effort to facilitate the licensing process. The study also revealed that the collision risk should be assessed based on distribution and spatial usage data and not just the distances between wind farms and griffon vulture colonies or communal roosts. In fact, research on species foraging behaviour through field observations and telemetry schemes could reliably define its home range including all aggregation points with heavy vulture use. In this context our results pinpoint the need for distribution density and sensitivity maps and the setting up of minimum buffer zones around the griffon breeding/roosting sites as elsewhere in the EU [20, 52, 60–63]. Apart from the species core areas, sensitivity mapping should consider certain landscape features that are used by soaring vultures in order to gain height and disperse. All these sites when superimposed on wind capacity maps could highlight potential "high-risk" areas where foraging griffons are most likely to be threatened by wind farm projects. Flight paths can also be identified by aerodynamic models combined with mapping of wind currents and thermal updraft producing spots (e.g. ridges, rocky outcrops, gorges, peninsulas, etc.). Unfortunately, no sensitivity mapping has been elaborated at national scale, or any spatial biodiversity data (e.g. nests, roosts) have been adequately buffered so to minimize wind farm impacts. In addition, dispersal or nursery areas and flight pathways have not been identified and delineated. Proper site planning during project development, curtailment of problematic turbines during critical periods and improved assessments to predict impacts on wildlife are still needed.

Overall there are a lot of concluding remarks on the options for reducing wildlife impacts, namely, avoidance when planning, minimization while designing, reduction at construction, compensation during operation and restoration as part of decommissioning. A mitigation measure that is being currently suggested in Greece and statutory fortified is using radars for implementing shutdown of wind turbines on demand [13]. This technique has been regarded as the most promising one since griffon mortality at some wind farms has dropped by 50% with an energy reduction of less than 1% by the short stops of wind turbines [13]. However, its efficacy is quite questionable in the case of griffons in Crete as this technique might be useless in areas with constant vulture activity almost throughout the day and the year (e.g. breeding and roosting cliffs, village and stockyards' garbage dumps, supplementary feeding stations, etc.) For this reason we argue that proper site planning should be prioritized over mitigation techniques during the elaboration of impact assessment studies. This is particularly so considering that predicted fatalities are often significantly lower than the recorded ones, suggesting that the factors affecting collision rates are often missed or not properly evaluated during the methodological procedure of risk assessment [8].

Collisions to wind turbines can be an unsustainable source of mortality for species such as vultures with slow population growth rates [17, 52, 64, 65]. This is especially so for griffons as wind turbines might act synergistically with other man-induced negative factors such as the illegal use of poisoned baits against mammalian carnivores or low availability of livestock carrion [66, 67]. Collisions at wind turbines should be assessed in relation to the aforementioned factors as their impact could accelerate the collapse of certain colonies near the most influential wind farms. The estimated annual population loss (11%) in this study might gradually decrease as birds habituate and avoid wind turbines. In any case wind farm polygons will still act as "hotspot" for the species, and the impact on its population will depend greatly on the age group mostly affected. A reduction of the survival of the species rates should be projected in the long run involving various age-dependent population viability analyses. In addition, the impact of the estimated casualties should be considered on the favorable reference values (FRVs) at the population level in the NATURA 2000 sites the species occurs [68].

Based on the outcome of the current study, we propose that the most effective action for minimizing wind farm impact on birds would be proper site planning that avoids the construction of wind energy infrastructure in certain areas. The NATURA 2000 sites should be prioritized as shown in the present case where the percent change of fatalities dropped by over twofold by excluding them from wind farming. Moreover under this scenario (#2), the wind energy facilities that would remain active or under construction on the island would number ca. 255 wind farms which could potentially generate more than 2500 MW. In the same context, it is worth mentioning that even if all NATURA 2000 sites in Europe and other areas designated for nature protection were theoretically excluded from wind energy production, there would still be enough wind energy available to supply three to seven times the total estimated energy demand in 2020 and 2030 [69]; the development of wind farms could be achieved with the minimal expense on wildlife. Furthermore the European Environment Agency has calculated that the technical potential for onshore wind energy in Europe is over 10 times total electricity consumption, and that excluding NATURA 2000 and other protected areas would reduce this by only 13.7% [69]. In Crete, a relevant study revealed that even by setting aside the NATURA 2000 site network and buffering sensitive areas (e.g. archaeological sites, biodiversity hotspots, urban settlements), there would be still enough space onshore for the positioning of more than 4000 MW [70]. Nevertheless, as wind farms are land-intensive and incorporate many accompanying works (e.g. construction of electricity grids and road networks), they would result in habitat encroachment of previously inaccessible areas. These interventions could produce irreversible damage on the quality of the vultures' foraging habitat by turning upland pastures into energy production areas. In the long term, the latter threat might have a more significant impact on the conservation of vultures on Crete than their direct mortality on wind turbine blades.

Appendix

Description of the parameters used in the collision risk model for estimating griffon vulture fatalities in wind energy facilities in Crete

Input parameters in the collision risk model (CRM)	CRM functions and parameter abbreviations	CRM fixed values	
Griffon vulture home range			
Pixel ID number	π_i		
Pixel length	α	5 km	
Pixel area	A_T	25 km^2	
Foraging daily time budget	f_t	6.8 h	8.4 h
Flight speed	f_v	43 km/h	
Total foraging distance	$f_d = f_t * f_v$	292.4 km	361.2 km
Colony home range	h	692 km^2	
No home range pixels	$h_p = h/A_T$	~28 pixels	
No pixels visited per foraging sortie per day	$n_p = f_d/\alpha$	58.5	72.2
No visits/home range pixel per day	$R_{pc} = n_p/h_p$	2	3
Flight height (median)	H	190 m	
Risky flights correction function	$Cf = (h/H) * (a_w/A_T)$		
Winter (days)	N_w	150	
Summer (days)	N_s	210	
No individuals per pixel (π_i)	n_i		
Population size of colony j	N_j		
No griffons using home range of colony j	Nh_j		
Relative presence of griffon of colony j in pixel i	$P_i = n_i / Nh_j$		
Collision probability	P_{col}		

(continued)

Input parameters in the collision risk model (CRM)	CRM functions and parameter abbreviations	CRM fixed values
Avoidance probability	Av	75%
No fatalities per day per griffon in its home range	$Q_w = [R_{pc}*Cf*P_i*P_{col}*(1-Av)]$	
No of fatal flights per day per colony	$D_j = N_j * Q$	
No fatalities per colony in winter	$Dw_j = D_j * N_w$	
No fatalities per colony in summer	$Ds_j = D_j * N_s$	
Cumulative griffon fatalities per year	$D_{tot} = Dw_j + Ds_j$	
Griffon vulture biometrics in Crete		
Length (from head to tail)	l	1.1 m
Wing span	WS	2.56 m
Reduction function of wingspan	WSrf	0.676
Flight type (0, flapping; 1, gliding)	F/G	1
Wind farm and turbine specifications		
Wind turbine height	h	45–85
Wind farm polygon area	a_w	
Number of blades	No Blades	3
Maximum length of blade chords	MaxChord	2.3–4 m
Rotor diameter	D	33–112
Rotation speed (sec)	RotPeriod	
Rotor pitch	2R/3	

Values in void cells had to be calculated or set according to technical specifications of wind turbine type. Double values refer to winter and summer periods

References

1. Huntley, B., Collingham, Y.C., Green, R.E., Hilton, G.M., Rahbeck, C., Willis, S.G.: Potential impacts of climatic change upon geographical distribution of birds. Ibis. **148**, 8–28 (2006)
2. EC (European Commission): Directive 2001/77/EC of the European Parliament and of the Council of 27 September 2001 on the promotion of electricity produced from renewable energy sources in the internal electricity market (Document L0077-20040501) (2001)
3. Nelson, H.K., Curry, R.C.: Assessing avian interactions with wind plant development and operation. Translations 60th North American Wildlife and Nature Resource Conferences. In: Wildlife Management Institute (ed.), Minneapolis, pp. 266–287 (1995)
4. Bevanger, K.1998: Biological and conservation aspects of bird mortality caused by electricity power lines: a review. Biol. Conserv. **86**, 67–76 (2004)
5. Ratcliffe, D.A.: Bird Life of Mountain and Upland. Cambridge University Press, Cambridge (1990)
6. Smallwood, K.S., Karas, B.: Avian and bat fatality rates at old-generation and repowered wind turbines in California. J. Wildl. Manag. **73**, 1062–1071 (2009)
7. Strickland, M.D., Arnett, E.B., Erickson, W.P., Johnson, D.H., Johnson, G.D., Morrison, M.L., Shaffer, J.A., Warren-Hicks, W.: Comprehensive Guide to Studying Wind Energy/Wildlife Interactions. Prepared for the National Wind Coordinating Collaborative, Washington, DC, USA (2011)

8. Ferrer, M., de Lucas, M., Janss, G.F.E., Casado, E., Muñoz, A.R., Bechard, M.J., Calabuig, C.P.: Weak relationship between risk assessment studies and recorded mortality in wind farms. J. Appl. Ecol. **49**, 38–46 (2012)
9. Hötker, H., Thomsen, K.-M., Jeromin, H.: Impacts on Biodiversity of Exploitation of Renewable Energy Sources: The Example of Birds and Bats-Facts, Gaps in Knowledge, Demands for Further Research, and Ornithological Guidelines for the Development of Renewable Energy Exploitation. Michael-Otto-Institutim NABU, Bergenhusen (2006)
10. Langston, R.H.W., Pullan, J.D.: Wind Farms and Birds: An Analysis of the Effects of Wind Farms on Birds, and. Guidance on Environmental Assessment Criteria and Site Selection Issues Report by Birdlife International on behalf of the Bern Convention. Sandy, RSPB, Sandy (2003)
11. Sovacool, B.K.: Contextualizing avian mortality: a preliminary appraisal of bird and bat fatalities from wind, fossil-fuel, and nuclear electricity. Energy Policy. **37**, 2241–2248 (2009)
12. Barrios, L., Rodriguez, A.: Behavioural and environmental correlates of soaring-bird mortality at on-shore wind turbines. J. Appl. Ecol. **41**, 72–81 (2004)
13. Birdlife International: Review and Guidance on use of "shutdown-on-demand" for Wind Turbines to Conserve Migrating Soaring Birds in the Rift Valley/Red Sea Flyway. Regional Flyway Facility, Amman (2015)
14. Dahl, E.L., May, R., Hoel, P.L., Bevanger, K., Pedersen, H.C., Røskaft, E., Stokke, B.G.: White-tailed eagles (Haliaeetus albicilla) at the Smøla wind-power plant, Central Norway, lack behavioural flight responses to wind turbines. Wildl. Soc. Bull. **37**, 66–74 (2013)
15. Drewitt, A.L., Langston, R.H.W.: Assessing the impacts of wind farms on birds. Ibis. **148**, 29–42 (2006)
16. Smallwood, K.S., Thelander, C.G.: Developing Methods to Reduce Bird Mortality in the Altamont Pass Wind Resource Area. Final Report by to the California Energy Commission. Public Interest Energy Research-Environmental Area, Contract no. 500 – 01– 019. BioResource Consultants, California (2004)
17. Vasilakis, D.P., Whitfield, D.P., Schindler, S., Poirazidis, K.S., Kati, V.: Reconciling endangered species conservation with wind farm development: Cinereous vultures (*Aegypius monachus*) in south-eastern Europe. Bio. Conserv. **196**, 10–17 (2016)
18. De Lucas, M., Janss, G.F.E., Whitfield, D.P., Ferrer, M.: Collision fatality of raptors in wind farms does not depend on raptor abundance. J. Appl. Ecol. **45**, 1695–1703 (2008)
19. Hunt, W.G., Jackman, R.E., Hunt, T.L., Driscoll, D.E., Culp, L.: A Population Study of Golden Eagles in the Altamont Pass Wind Resource Area: Population Trend Analysis 1994–1997 Report to National Renewable Energy Laboratory, Subcontract XAT-6-16459-01. Predatory Bird Research Group, University of California, Santa Cruz (1999)
20. De Lucas, M., Ferrer, M., Bechard, M.J., Muñoz, A.R.: Griffon vulture mortality at wind farms in southern Spain: distribution of fatalities and active mitigation measures. Biol. Conserv. **147**, 184–189 (2012)
21. Erickson, W., Johnson, G., Young, D., Strickland, D., Good, R., Bourassa, M., Bay, K., Sernka, K.: Synthesis and Comparison of Baseline Avian and Bat Use, Raptor Nesting and Mortality Information from Proposed and Existing Wind Developments. Final report by WEST Inc. prepared for Bonneville Power Administration, Portland, Oregon (2002)
22. Hoover, S.I., Morrison, M.L.: Behavior of Red-tailed Hawks in a wind turbine development. J. Wildl. Manag. **69**, 150–159 (2005)
23. Johnson, G.D., Young, D.P., Erickson, W.P., Clayton, E., Derby, C.E.M., Dale Strickland, M.D., Good, R.E.: Wildlife Monitoring Studies Seawest Windpower Project, Carbon County, Wyoming (2000a)
24. Johnson, G.D., Erickson, W.P., Strickland, M.D., Shepherd, M.F., Shepherd, D.A.: Avian Monitoring Studies at the Buffalo Ridge Wind Resource Area, Minnesota: Results of a 4-Year Study. Technical report by WEST Inc. prepared for Northern States Power Co., Minneapolis (2000b)
25. EC (European Commission): Directive 2001/42/EC of the European Parliament and of the Council of 27 June 2001 on the assessment of the effects of certain plans and programmes on the environment (Document 32009 L0147) (2001)

26. EC (European Commission): Directive 2011/92/EU of the European Parliament and of the Council of 13 December 2011 on the assessment of the effects of certain public and private projects on the environment (Document 32011 L0092) (2011)
27. EC (European Commission): Wind Energy Developments and NATURA 2000. Guidance document. (2011). https://doi.org/10.2779/98894
28. EC (European Commission): Directive 2009/147/EC of the European Parliament and of the Council of 30 November 2009 on the conservation of wild birds (Document 32001 L0042) (2009)
29. EC (European Commission): Directive 92/43/EC of the European Parliament and of the Council of 21 May 2009 on the Conservation of natural habitats and of wild fauna and flora (Document 31992L0043) (1992)
30. Dimalexis, A., Kastritis, T., Manolopoulos, A., Korbeti, M., Fric, J., Saravia Mullin, V., Xirouchakis, S., Bousbouras, D.: Identification and Mapping of Sensitive Bird Areas to Wind Farm Development in Greece, p. 126. Hellenic Ornithological Society, Athens (2010)
31. Vasilakis, D.P., Whitfield, D.P., Kati, V.: A balanced solution to the cumulative threat of industrialized wind farm development on cinereous vultures (*Aegypius monachus*) in south-eastern Europe. PLoS One. **2**(2), e0172685 (2017). https://doi.org/10.1371/journal.pone.0172685
32. Handrinos, G., Akriotis, T.: The Birds of Greece. C. Helm, A & C. Black, London (1997)
33. Xirouchakis, S.M., Tsiakiris, R.: Status and population trends of vultures in Greece. In: Donázar, J.A., Margalida, A. & Gampion, A. (eds.) Vultures Feeding Stations & Sanitary Legislation: A Conflict and Its Consequences from the Perspective of Conservation Biology. Munibe (*suppl*) (29), pp. 154–171 (2009)
34. Grove, A.T., Mooney, J., Rackham, O.: Crete and the South Aegean Islands: Effects of Changing Climate on the Environment. European Community, p. 439, Brussels (1991)
35. Handrinos, G.: The status of vultures in Greece. In: Newton, I., Chancellor, R.D. (eds.) Conservation Studies in Raptors ICBP Technical Publication, No 5. Salonica, pp. 103–115. WWGBP, Berlin (1985)
36. Bourdakis, S., Alivizatos, H., Azmanis, P., Hallmann, B., Panayotopoulou, M., Papakonstantinou, C., Probonas, N., Rousopoulos, Y., Skartsi, D., Stara, K., Tsiakiris, R., Xirouchakis, S.: The situation of Griffon Vulture in Greece. In: Slotta-Bachmayr, L., Bögel, R., Camina, C.A. (eds.) The Eurasian Griffon Vulture (*Gyps fulvus*) in Europe and the Mediterranean. Status Report and Action Plan, pp. 48–56. EGVWG, Vienna (2004)
37. Legakis, A., Maragou, P. (eds.): Red Data Book of Endangered Animals in Greece, pp. 403–405. Hellenic Zoological Society, Athens (2009). in Greek
38. Xirouchakis, S.M., Mylonas, M.: Griffon Vulture (*Gyps fulvus*) distribution and density in Crete. Isr. J. Zool. **50**, 341–354 (2004)
39. Eichhorn, M., Johst, K., Seppelt, R., Drechsler, M.: Model-based estimation of collision risks of predatory birds with wind. Ecol. Soc. **17**, 1 (2012)
40. Band, W.: Windfarms and Birds: Calculating a Theoretical Collision Risk Assuming No Avoidance Guidance notes series, p. 10. Scottish Natural Heritage, Inverness (2000)
41. Band, W., Madders, M., Whitfield, D.P.: Developing field and analytical methods to assess avian collision risk at wind farms. In: de Lucas, M., Janss, G., Ferrer, M. (eds.) Birds and Wind Farms, Risk Assessment and Mitigation, pp. 259–275. Servicios Informativos Ambientales/Quercus, Barcelona (2005)
42. Chamberlain, D.E., Rehfisch, M.R., Fox, A.D., Desholm, M., Anthony, S.J.: The effect of avoidance rates on bird mortality predictions made by wind turbine collision risk models. Ibis. **148**, 198–202 (2006)
43. Smales, I., Muir, S., Meredith, C., Baird, R.: A description of the biosis model to assess risk of bird collisions with wind turbines. Wildl. Soc. Bull. **37**, 59–65 (2013)
44. RAE (Regulatory Authority for Energy): Geoinformational map of Greek Regulatory authority for energy. Available from http://www.rae.gr/geo/?lang=EN. (2015)
45. Xirouchakis, S.M., Andreou, G.: Foraging behaviour and flight characteristics of griffon vultures (*Gyps fulvus*) on the island of Crete (Greece). Wildl. Biol. **15**, 37–52 (2009)

46. Madders, M., Whitfield, D.P.: Upland raptors and the assessment of wind-farm impacts. Ibis. **148**, 43–56 (2006)
47. Xirouchakis S.M.: Monitoring the impact on the avifauna of the operation of the wind farm of TERNA-ENERGEIAKH at the site of "Perdikorifi" of Municipal District of Agia Varvara, Heraklion prefecture. Unpublished report, p. 76. Natural History Museum of Crete, University of Crete, Heraklion (2009)
48. Xirouchakis, S.M., Poulakakis, N.: Biometrics, sexual dimorphism and gender determination of griffon vultures (*Gyps fulvus*) from Crete. Ardea. **96**, 91–98 (2008)
49. Zar, J.H.: Biostatistics, 3rd edn. Prentice-Hall, Inc, Upper Saddle River (1996)
50. Percival, S.M.: Birds and wind turbines in Britain. Brit. Wildl. **12**, 8–15 (2000)
51. Telleria, J.L.: Overlap between wind power plants and griffon vultures *Gyps fulvus* in Spain. Bird Study. **56**, 268–271 (2009)
52. Carrete, M., Sánchez-Zapata, J.A., Benítez, J.R., Lobón, M., Montoya, F., Donázar, J.A.: Mortality at wind-farms is positively related to large-scale distribution and aggregation in griffon vultures. Biol. Conserv. **145**, 102–108 (2012)
53. Gill, J.P., Sales, D., Pullinger, M., Durward, J.: The Potential Ornithological Impact of the Proposed Kentish Flats Offshore Wind farm. Ornithological Technical Addendum Report to GREP UK Marine Ltd. UK: Environmentally Sustainable Systems, Edinburgh (2002)
54. SNH (Scottish Nature Heritage): Strategic Locational Guidance for Onshore Wind Farms in Respect to the Natural Heritage Policy Statement No. 02/02, update March 2009. Scottish Natural Heritage, Inverness (2009)
55. Cook, A.S.C.P., Burton, N.H.K., Humphreys, E.M., Masden, E.A.: The Avoidance Rates of Collision Between Birds and Offshore Turbines. In: S.M.a.F (ed.) Science, p. 247. Edinburgh (2014)
56. Xirouchakis S.M., Andreou, G., Armeni, E., Probonas, M.: Environmental Monitoring Program with Emphasis on the Avifauna at the Wind Energy Station at the Site of "Agios Kyrillos", Municipal District of Gortyna, Heraklion prefecture. Unpublished report, pp. 82. Natural History Museum of Crete, University of Crete, Heraklion (2014)
57. Xirouchakis S.M., Andreou, G., Probonas, M.: Environmental Monitoring Program with Emphasis on the Avifauna at the Wind Energy Station at the Site of "Koutsotroulos", Municipality of Mylopotapos, Rethymnon prefecture. Unpublished report, p. 64. Natural History Museum of Crete, University of Crete, Heraklion (2017)
58. Martin, G.R.: Understanding bird collisions with man-made objects: a sensory ecology approach. Ibis. **153**, 239–254 (2011)
59. Northrup, J.M., Wittemyer, G.: Characterising the impacts of emerging energy development on wildlife, with an eye towards mitigation. Ecol. Lett. **16**, 112–125 (2013)
60. Bright, J.A., Langston, R.W., Bullman, R., Evans, R.J., Gardner, S., Pearce-Higgins, J., Wilson, E.: Bird Sensitivity Map to Provide Locational Guidance for Onshore Wind Farms in Scotland. RSPB Research Report No 20, Sandy Bedfordshire (2006)
61. Bright, J.A., Langston, R.W., Bullman, R., Evans, R.J., Gardner, S., Pearce-Higgins, J.: Map of bird sensitivities to wind farms in Scotland: a tool to aid planning and conservation. Biol. Conserv. **141**, 2342–2356 (2008)
62. Fielding, A.H., Whitfield, D.P., McLeod, D.R.A.: Spatial association as an indicator of the potential for future interactions between wind energy developments and golden eagles *Aquila chrysaetos* in Scotland. Biol. Conserv. **131**, 359–369 (2006)
63. Walker, D., Mcgrady, M., McLuskie, A., Madders, M., McLeod, D.R.A.: Resident Golden Eagle ranging behaviour before and after construction of a windfarm in Argyll. Scottish Birds. **25**, 24–40 (2005)
64. Carrete, M., Sánchez-Zapata, J.A., Benítez, J.R., Lobón, M., Donázar, J.A.: Large scale risk-assessment of wind-farms on population viability of a globally endangered long-lived raptor. Biol. Conserv. **142**, 2954–2961 (2009)
65. Sanz-Aguilar, A., Sánchez-Zapata, J.A., Carrete, M., Benítez, J.R., Ávila, E., Arenas, R., Donázar, J.A.: Action on multiple fronts, illegal poisoning and wind farm planning, is required to reverse the decline of the Egyptian vulture in Southern Spain. Biol. Conserv. **187**, 10–18 (2015)

66. Cortés-Avizanda, A., Blanco, G., DeVault, T.L., Markandya, A., Virani, M.Z., Brandt, J., Donázar, J.A.: Supplementary feeding and endangered avian scavengers: benefits, caveats, and controversies. Front. Ecol. Environ. **14**, 191–199 (2016)
67. Margalida, A., Colomer, M.A.: Modelling the effects of sanitary policies on European vulture conservation. Sci. Rep. **2**, 753 (2012)
68. Evans, D., Arvela, M.: Assessment and Reporting Under Article 17 of the Habitats Directive Explanatory Notes & Guidelines for the Period 2007–2012, p. 123. European Topic Centre on Biological Diversity, Brussels (2011)
69. EEA (European Environmental Agency): Europe's Onshore and Offshore Wind Energy Potential. An Assessment of Environmental and Economic Constraints. – EEA Technical report No 6/2009 (2009)
70. Tsoutsos, T., Kokologos, D., Tsitoura, M., Vasilomichelaki, A., Xirouchakis, S., Probonas, M., Nikolopoulou, S.: Special Study for the Site Planning and Sustainable Development of Wind Farms in the Region of Crete, p. 172. Crete Polytechnic School, Chania (2011)

Printed in the United States
By Bookmasters